BAD NATURE

BAD NATURE

HOW RAT CONTROL SHAPES
HUMAN AND NONHUMAN WORLDS

ANDREW McCUMBER

The University of Chicago Press
Chicago and London

The University of Chicago Press, Chicago 60637
The University of Chicago Press, Ltd., London
© 2025 by The University of Chicago
All rights reserved. No part of this book may be used or reproduced in any manner whatsoever without written permission, except in the case of brief quotations in critical articles and reviews. For more information, contact the University of Chicago Press, 1427 E. 60th St., Chicago, IL 60637.
Published 2025
Printed in the United States of America

34 33 32 31 30 29 28 27 26 25   1 2 3 4 5

ISBN-13: 978-0-226-83896-0 (cloth)
ISBN-13: 978-0-226-83898-4 (paper)
ISBN-13: 978-0-226-83897-7 (e-book)
DOI: https://doi.org/10.7208/chicago/9780226838977.001.0001

Library of Congress Cataloging-in-Publication Data

Names: McCumber, Andrew Hammond, author.
Title: Bad nature : how rat control shapes human and nonhuman worlds / Andrew McCumber.
Other titles: How rat control shapes human and nonhuman worlds
Description: Chicago : The University of Chicago Press, 2025. | Includes bibliographical references and index.
Identifiers: LCCN 2024046663 | ISBN 9780226838960 (cloth) | ISBN 9780226838984 (paperback) | ISBN 9780226838977 (ebook)
Subjects: LCSH: Rats—Social aspects—North America. | Rats—Social aspects—California—Los Angeles. | Rats—Control—California—Los Angeles. | Rats—Social aspects—Alberta. | Rats—Control—Alberta. | Rats—Social aspects—Galapagos Islands. | Rats—Control—Galapagos Islands. | Rats—Control—Social aspects. | Human-animal relationships
Classification: LCC SB994.R2 M44 2025 | DDC 363.7/8—dc23/eng/20241128
LC record available at https://lccn.loc.gov/2024046663

♾ This paper meets the requirements of ANSI/NISO Z39.48-1992 (Permanence of Paper).

"Rat Song" from *Selected Poems: 1965–1975* by Margaret Atwood. Copyright © 1976, renewed 2004 by Margaret Atwood. Used by permission of HarperCollins Publishers.

FOR LAURA

You think: *That one's too clever*,
she's dangerous, because
I don't stick around to be slaughtered
and you think I'm ugly too
despite my fur and pretty teeth
and my six nipples and snake tail.
All I want is love, you stupid
humanist. See if you can.

—MARGARET ATWOOD, "RAT SONG"

Okay, everybody tuck your pants into your socks.

—MOE SZYSLAK, *THE SIMPSONS*
(in response to a torrent of rats entering his bar)

CONTENTS

1 Introduction 1

PART I ANIMALS IN BOUNDARY WORK

2 Rats and "Boundary Work" on the Canadian Prairie 31

3 "You Can't Ignore the Rat": Guarding Alberta's Moral Character 51

PART II THE BORDERS OF URBAN NATURE

4 Rats and the Indoors/Outdoors Divide 79

5 Bulky Items: The "Rat Problem" and the "Homeless Problem" in Downtown LA 99

PART III ECOLOGIES OF MEANING

6 Ecologies of Meaning in Ecuador's Galápagos Islands 127

7 Killing for Life: Morally Acceptable Lives and Deaths in Environmental Conservation 153

8 Conclusion 179

Acknowledgments 191
Appendix A: Regression Results Table 195
Appendix B: Natural Language Processing 196
References 203 *Index* 217

× one ×

INTRODUCTION

In the fall of 2016, amid the depths of an unrelated Google image search, I stumbled across a graphic created by an online retailer of pest control products called Do My Own Pest Control. The image was a map that purported to display the "Global Rat Distribution," and was color coded in a binary red-and-blue scheme representing areas where rats live and areas where they do not. The Earth's landmass, as depicted in this Robinson projection, is nearly universally cherry red, indicating a world thoroughly conquered by these small creatures. The only exceptions were the polar regions beyond the Antarctic and Arctic Circles and a curious blue puzzle piece within the vast red expanse of North America that would send me down the path toward writing this book. The mostly rectilinear boundaries of this blue, rat-free shape were those of the Canadian province of Alberta. Alberta, I soon learned, had a long-standing government program dedicated to ensuring that the territory within that blue puzzle piece remained free of rats.

As a trained social scientist, of course, I knew to take the geographic pronouncements of a map produced by a company hawking ant spray and poison bait pellets as something less than the result of a rigorous scientific study. On the other hand, as a sociologist studying cultural meaning, I was instantly fascinated by the narrative this map told of a rat-free Canadian province and its stark borders separating it from the domain of rats. Later, to conduct fieldwork for this project, I road-tripped to Alberta from my home in Santa

Barbara, California. In that long drive spread across several days, I frequently thought about the landscape I was traversing in terms of that map, while I traveled toward its conspicuous blue puzzle piece. All of it—the California coast, the Cascade Mountains, the tan, monochromatic expanse of eastern Washington—was supposedly rat territory. On the last day of my drive, after several scenic hours twisting and turning through the Idaho panhandle and the Canadian Rockies, I reached the Great Continental Divide, which provides the only one of Alberta's boundaries that is not an artificially straight line of longitude or latitude. As I approached a sign welcoming me to Alberta, I recall scanning the sides of the road, as if I might see some sort of evidence of the province's rat-free distinction like a teeming population of rodents gathered at the border, kept on the British Columbia side by some invisible force field.

As it turns out, there is no rat-proof fence along the border. Nor, to the disappointment of one colleague who humorously speculated so, is there a fiery moat of gasoline lining the 110th meridian west, which forms Alberta's border with Saskatchewan to the east. Nevertheless, I found that the practice of rat control itself is deeply connected to boundaries, both spatial and symbolic. In what follows, I explore how managing the life and death of rats brings order and clarity to these meaningful distinctions and what cultural implications this work has.

In a broader sense, this is a story about our human social relationships to nonhuman others. Both in spite of the generally negative feelings many people harbor for them and also precisely because of them, rats are perfect protagonists for an exploration of these human-nonhuman relationships. Very shortly after deciding to pursue this research project, I learned that everyone, it seems, has a rat story. I have heard about rats who nested inside a car engine, adding an unpleasant wrinkle to one colleague's daily commute from their home on a mountain road. On one occasion I met up with a friend on the night he returned home from a two-month trip away from home, and he informed me that he was greeted upon his arrival by a deceased "toilet rat" that had evidently swum up through his sewage pipes. When we spoke, he was unsure what the end of that unpleasant story would be, as he told me he had lowered the lid, flushed, and simply left the house to meet me, unable to bear checking whether he had succeeded in disposing of the corpse. (He had.) A fellow sociologist had a peaceful retreat with their spouse spoiled when a particularly brazen rat scurried across the floor of their vacation rental, forestalling a peaceful night's sleep until they were sure they had removed it. Most amusingly, perhaps, I heard of an eccentric uncle who kept a BB gun by his bathtub, in case the chance arose for him to hunt the rodents while enjoying a soak.

Two important things about rats are highlighted by these anecdotal data. First, rats are uniquely intertwined in the social fabric of human life. As the Do My Own Pest Control map described above insists, rats live nearly everywhere in the world. Moreover, they thrive perhaps nowhere more than in our shadows, readily taking advantage of the food sources and warm environments created by human social life. Second, the emotional responses that our encounters with these animals elicit tend to have a negative valence. After all, with some help from Charles Schulz's *Peanuts* comic strip, "Rats!" became a common exclamation of disgust or disappointment sometime in the middle of the twentieth century. As I will return to later, rats are not universally reviled, as some like to keep them as pets (in the interest of full disclosure, I myself even had pet rats as a child, partially because of the restrictions on dogs and cats in the rental properties in which my family lived). Nonetheless, rats are most often met with some combination of revulsion and animus.

These two observations will serve as a jumping-off point to examine the cultural meaning of our explicitly violent and antagonistic relationships with nonhuman others. Though some multispecies scholars have explored these more contentious interspecies relationships (Timothy Pachirat's 2011 ethnography of industrial slaughterhouses being a prime example), they remain undertheorized compared to other genres of human-animal relationships characterized chiefly by affection, symbiosis, or harmony. We stand to gain a much more detailed account of the sociological importance of nonhumans by turning to relationships like those between humans and rats. Specifically, what are the social and cultural implications of the instances where human populations have endeavored to systematically exterminate another species? This book will examine this question in a multi-sited study of rat extermination, eradication, and control. It will draw primarily from ethnographic, interview, and textual data on three locations: Alberta, Canada; Los Angeles, California; and the Galápagos Islands of Ecuador. As already noted, the rural province of Alberta has claimed to be "rat-free" for decades on the strength of its government rat control program that monitors the border with Saskatchewan for rats. In Los Angeles, a rat infestation in City Hall sparked public outcry that metastasized into a symbolic referendum on nearby homeless encampments. Finally, in the Galápagos, rats are among the "invasive species" that environmental conservation groups attempt to eradicate in order to protect native species habitats.

Through analysis of these three case studies, I advance two overarching arguments about the broader phenomenon of rat control and the cultural lessons it offers:

First, rat control is a social practice that draws and clarifies the boundaries of nature and society. On the surface, rat extermination is a project with a wide range of possible motivations. The stated rationale for the rat program in Alberta is an economic one, for instance, where keeping the province free of rats is seen as guarding against lost agricultural production. Elsewhere, rat control is a public health initiative concerned with preventing the spread of infectious diseases (LA) or part of an environmental conservation effort (Galápagos). Whatever their official intentions may be, however, these various rat extermination projects are all efforts to lend clarity to the spatial and symbolic boundaries between nature and society. In other words, by controlling the lives and movements of rats, these organizations manage the acceptable terms of our human coexistence with the nonhuman world: Albertan pest control officers enforce the appropriate environmental conditions that make for good, dignified farming; the LA city government guards public faith in the threshold between "indoors" and "outdoors"; and conservationists in the Galápagos work to prevent pristine wilderness from being defiled by human influence. Rats demand such nature/society boundary work because their existence defies this taken-for-granted distinction. They are nonhuman animals but are nonetheless far from charismatic wildlife. They bring the nonhuman world to our doorsteps, thriving along the margins of our population centers always in proximity to us, despite how much we may desire to cast them off to the hinterlands.

Second, rat control enforces a meaningful hierarchy of living things that mirrors and is entangled with social inequalities. Rat control programs reveal how, at its core, the work of managing the boundaries between human society and nature means adjudicating the right to life itself. Rats, I find, lie near the bottom of a symbolic hierarchy of species that extends from human life at the top, cherished beings like companion animals and majestic wildlife just below us, on down to the irksome pests and "invasive species" that call for systematic extermination. In this project's cases, this hierarchy mirrors specifically human systems of inequality and becomes enmeshed with them. Rats come to act as symbolic proxies for unwanted immigration in Alberta. The "rat problem" in LA's City Hall becomes connected to and even indistinguishable from the contentious politics of homeless policy. Local populations and NGOs clash over who has the legitimate right to effect environmental change and for what purposes on the Galápagos Islands.

Together, these findings demand that we extend the lessons of environmental justice, the notion that the burden of environmental problems falls disproportionately on already marginalized populations (Bullard 2018), to the cultural imaginations of nature and environmentalism itself. Managing our relationships with the nonhuman world, rat control shows us, is inherently an expression of power. Charting a just and sustainable future, therefore, will require that we reckon with the often uncomfortable question of what place the rats of the world will have in the worlds we envision.

A SOCIOLOGY OF RATS AND EXTERMINATION

This is a book about rats, and it is also a book about extermination. That much may be obvious by this point, but it bears being explicit about the intention behind organizing this project around these two themes. There is much that is sociologically interesting about rats that does not relate specifically to projects of rat extermination. It is not difficult to imagine a fascinating project within this broader "sociology of rats" that never leaves the confines of New York City's subway system, with its famously large population of murine denizens who occasionally become viral internet characters for snatching magnificently large slices of pizza and have even prompted the city to appoint a "rat czar."[1] Likewise, one might devote an entire book to the topic of extermination without paying any unique attention to rats compared to the many other beings that humans target with warfarin pellets, firearms, and other tools of lethal force. In *War and Nature*, for instance, historian Edmund Russell argues that the insecticide and chemical industries evolved in close connection with the military-industrial complex between World War I and the publication of Rachel Carson's *Silent Spring* (Russell 2001). All this is to say, by choosing these two organizing principles, my aim is not to tell the entire sociological story of human-rat relationships *on top of* the entire story of extermination. Rather, I choose these themes because I believe that we learn something unique and important about the cultural meaning of our relationships with nature and the nonhuman world by considering them together.

Far from two ultimately arbitrary topics, rats and extermination are mutual epitomes of each other. On one hand, the most salient emotional responses we tend to associate with rats are those of fear, revulsion, and dislike. Occasionally rats, real or fictional, capture the hearts of the masses (such as the "pizza rat" alluded to above or Remy the rat chef in the Pixar film *Ratatouille*), but these instances seem to occur *in spite of* rats' more common associations, as exceptions that prove the rule. After all, most viewers who were delighted by *Ratatouille* would nonetheless be horrified by the sight of a living, breathing rat in their own kitchen. On the whole, in the Western world, we have less affinity for rats than perhaps any other mammalian species, and accordingly we are uniquely willing to kill them.[2] Some, of course, enjoy keeping rats as pets, but the comparative scope of this practice is dwarfed by that of the industrial project of rat extermination. Moreover, we exterminate lots and lots of things, and have done so in various places around the world for centuries.

---

1   New York City's rats are the topic of Robert Sullivan's excellent book *Rats: Observations on the History and Habitat of the City's Most Unwanted Inhabitants* (Sullivan 2005).

2   We of course kill lots of animals on a mass scale for food, but few would say this killing is motivated by anywhere near the animosity that characterizes the human-rat relationship.

Rats, however, are the animals most like us that we exterminate on such a large scale. Rat extermination is the variety of this animal relationship where we are most compelled to contend with the killing that is so obviously front and center.

Animals and the Environment in Sociology

With this in mind, a primary reason that rats and extermination intersect in a sociologically important way is that rats occupy a liminal space on the border between nature and society. As nonhuman animals, they are comfortably categorized as part of the natural world, but they are also a quintessential facet of human civilization, thriving in densely populated concrete jungles and epitomizing urban life. This proximity to human social life makes rats compel us to acknowledge how nonhuman others are more than just window dressing adorning the periphery of the analytically significant social world.

Sociology has struggled to fully embrace this principle over the course of its existence, leading to the development of environmental sociology as a subdiscipline in the 1970s. This coincided with the rise of the modern environmental movement in the United States. Perhaps the most iconic image associated with this period of environmental activism is *The Blue Marble*, the photograph of Earth taken by the *Apollo 17* mission that was set against a deep cobalt background to become the unofficial flag of Earth Day. By depicting the globe as a singular object, the image compels us to confront the finite nature of the natural environment and the resources it provides. Meanwhile, early environmental sociologists likewise argued that their broader discipline needed to reckon with this principle, as sociology, they said, had been molded by the same delusion that human society was "exempt" from the limits imposed by the natural world (Catton and Dunlap 1978, 1980). Put another way, the ill-conceived notion that the Earth could offer an inexhaustible bounty presented a grave threat to the future sustainability of human life, but also, relatedly, it produced a faulty set of assumptions for academics trying to understand social life. This naive idea of "human exemptionalism," they argued, was partly responsible for Émile Durkheim's assertion that "social facts" could only be explained by other social facts (Durkheim 2014).

The pioneering environmental sociologists Catton and Dunlap thus called for a Kuhnian "paradigm shift" (1962) that would legitimately incorporate the physical environment as a central factor in explaining social life.

While many, if not most, environmental sociologists would likely tell you that their subfield has yet to be fully embraced by the sociological mainstream, its enormous growth since those early days is an indication of how much explanatory power would be missed without duly considering the environment. The variety of different forms that environmental sociology has taken also demonstrates, though, that bringing the natural world into sociological analysis can mean many different things. Durkheim's original formulation that Catton and Dunlap critiqued emphasized the distinction between the natural and the social partly as a way of carving out a place for sociology that distinguished it from biology. In this sense, nature is defined broadly as everything that is not explicitly social. Thus, for sociologists, considering the natural world might mean analyzing how geographical and biophysical elements of a landscape might shape social activity or political attitudes (Freudenburg, Frickel, and Gramling 1995; Freudenburg and Gramling 1993), how uneven access to material resources or the deleterious effects of industrial processes might exacerbate inequalities (Bell 2013; Bullard 2018; Mohai, Pellow, and Roberts 2009; Nixon 2011; Pellow and Nyseth Brehm 2013; Voyles 2015), or any number of different ways the social world intersects with the nonhuman world. In this context, nonhuman animals present their own genre of environmental sociology. Animals constitute nonhuman nature, but in the form of sentient, individual organisms, that humans relate to in unique ways. By crossing paths with humans in the course of acting on their own impulses, needs, and motivations, animals are active participants in social life in ways that are more visible than other, similar forms of "nonhuman agency" that analysts have extended to microbes (Latour 1988), mushrooms (Tsing 2015), or ocean currents (Law 1987). Accordingly, early sociologists of human-animal relationships sought to expand the epistemological boundaries of the discipline by extending personhood to animals and putting them on equal footing with humans as agentic participants in social processes (Irvine 2008; Sanders 2010; Sanders and Arluke 1993).

These efforts have certainly not been without controversy. In a 1993 article in *Sociological Quarterly*, Clinton Sanders and Arnold Arluke use feminist standpoint theory to argue that humans can legitimately parse the perspectives of animals in their proximity as "nonverbal others" (Sanders and Arluke 1993). In a response to this piece, Richard Hilbert points out that this notion would seem to imply that women's contributions were entirely superfluous to the development of feminist theory, since men could intuit their perspectives by virtue of their proximity (1994). Relatedly, a shortcoming of many sociological treatments of nonhuman animals, and one which this book's focus on rats and extermination addresses, is that they gravitate toward a fairly narrow subset of human-animal relationships, giving most of their attention to the most proximate kinds of companion animals and other mostly harmonious interspecies interactions. This stems in part from the fact that nature itself is most often moralized as inherently "good," inspiring attempts to reap the benefits of connecting with it (Angelo 2021; Bell 2018). Some more recent animal-focused work has bucked this trend, including Colin Jerolmack's excellent ethnography *The Global Pigeon*, which explicitly focuses on an animal widely seen as an urban pest (2013). A large part of Jerolmack's book, however, focuses on pigeon keepers, who relate to the birds not as pests but as the objects of animal husbandry.

A number of other sociologists have also recently turned to humans' enduring fascination with and affection for birds to better understand our social relationships to animals and, by extension, to nature. Both Stefan Bargheer and Hillary Angelo probe tensions within communities of bird enthusiasts past and present about whether killing birds to collect as specimens is an acceptable way to act on their affection for the animals. At stake in these analyses are broader questions about how humans ought to interact with the nature they love and care for. The history of ornithologists using rifles as the primary tools for their pastime, along with the backlash to this practice from other bird lovers, indicates how the use of lethal force on animals is fraught terrain in the negotiation of what makes nature valuable. The opposition to killing in this instance indicates an underlying notion of nature, as embodied in animals like birds, as inherently good. Indeed, even the defenders of the practice share a passion and enthusiasm for the animals in question. In contrast to the topic of this book, the interspecies violence that Bargheer and Angelo consider is not motivated by animus.

Another recent sociological treatment of bird-watching, though, Elizabeth Cherry's ethnography *For the Birds*, emphasizes how not all birds are apt embodiments of an idealized nature. She notes how her participants distinguish between "good" and "bad" birds, the latter of which might be categorized as such because they have been deemed an "invasive" species causing habitat destruction or other perceived environmental harms. Bad birds run counter to the values that nature-loving bird-watchers bring to their hobby.[3] Cherry's typology of good and bad birds draws on Arluke and Sanders's (1996) related concept of a "sociozoologic scale," or an implicit categorization into which humans organize other nonhuman animals according to their conformity to acceptable roles in human social life. Arluke and Sanders propose "good" and "bad" categories of animals, where companion animals and other "useful" animals like livestock are considered good while pests and dangerous animals are deemed bad.

Cherry's use of this concept to describe good and bad birds specifically connects the meaning these animals carry with the environmental consciousness of their observers, which highlights the importance of ideas of nature to such a moralized hierarchy. The connections between the cultural meaning of animals and a generalized cultural idea of nonhuman nature add revealing texture to the notion of a "sociozoologic scale" and the value-laden categories that make it up. Our ideas of nature, what conditions make it "good," and what our human relationships to it ought to look like all underlie the meanings attributed to nonhuman animals. To put this another way, while the sociozoologic scale usefully describes how animals are grouped into "good" and "bad" animals, I argue that the underlying notion of "nature" itself as a cultural idea imbued with morally

---

3   Cherry's categorization recalls an insight of Jerolmack's that the extent to which animals are incorporated into the fabric of human social activity depends on their ability to perform "socially appropriate roles." This notion is echoed in Grazian's (2015) observations regarding the performative staging of enclosures and sensitivity to audience expectations in his ethnography of zoos.

inflected meaning is crucial for understanding categorizations like these. That nature is normatively conceptualized as "good," as well as being defined in opposition to humanity, adds an additional layer of complexity to the categorization of animal others. Partly to emphasize the importance of this objectified nature to our relationships with nonhuman animals, I shift the analytical focus from good and bad animals to good and bad nature. These latter terms capture the nonhuman world's capacity to support or contravene collective social goals. The moralized hierarchy that animals are ordered along is part of a broader cultural project of curating the category of good nature. Moreover, rats' ambiguous naturalness or unnaturalness (stemming from their proximity to human society and our own ambivalence or disdain for that proximity) makes them useful analytical subjects for better understanding this phenomenon.

Importantly, animals do symbolic work that contributes to drawing boundaries between good and bad nature whether or not the meaning-making that structures our interactions with them is obviously and explicitly tied to environmental values and ethics, as it is in the case of bird-watching. The symbolic work that animals perform also transcends human-animal relationships characterized by affection, wonder, or reverence and others characterized by animosity, resentment, or disgust. This book implores us to consider the notion that when we systematically exterminate another species, we do it not because that species serves no "socially appropriate role," but rather because they do in fact serve a very specific role that is evidenced by that extermination itself. The animosity and violence of interspecies extermination, in other words, is a meaningful process that clarifies the lines between good and bad nature.

Animals as Cultural Objects

By examining rat control through this lens, this book considers the role of rats (and, by extension, other animals) as cultural objects. To use Griswold's (1986) enduring definition, this means considering how rats comprise "shared significance embodied in form." In the sociological spirit of "making the familiar strange" (Mills 2000), the phenomenon of rat control appears so widespread that it is mundane, yet the extraordinary effort and resources that are marshaled for the task of exterminating animal others ought to compel us to probe the meaning of this practice and the creatures at its center. In other words, rats are not value-neutral elements of the biophysical backdrop that frames the social world. For us to become collectively inured to the killing of these animals has required cultural meaning-making that defines rats themselves as killable, or worse, as actively antagonistic enemies of humans.

Making sense of rat control, then, requires not just examining the physical practice of exterminating rats or the organizational approaches to that task (though these are both important), but also probing what rats mean to us in a symbolic sense and, by extension, what killing them means for the cultural conflicts that undergird environmental ones (Farrell 2015; Scoville 2022). By consolidating meaning into a discrete physical form, cultural objects occupy a duality of the discursive and the material. One expression of this duality is the distinction between "types," or broad schematic categories, and "tokens," or discrete manifestations of those categories (McDonnell 2023; Taylor, Stoltz, and McDonnell 2019). McDonnell illustrates these concepts with this example: "The type 'gun' may have many diverse tokens in the world: AR-19s, hunting rifles, nerf guns, clay pigeon shotguns, sniper rifles, guns in a video game, toy guns with red tips, or 3D-printed guns. Each new token we confront in the world requires us to reflect on and refine our fuzzy understanding of what counts as 'gun' as type" (McDonnell 2023).

Animals epitomize this same duality. Our relationship to the animal world is based on a set of categorizations, including biological taxonomy and more informal categories, which divide animals into "types." Historically, the formal categorization of animals and the assignment of moral qualities and other cultural meanings to them have been blended, as epitomized by the bestiaries of medieval Europe that combined encyclopedic descriptions of animals with didactic prose about their perceived virtues or moral failings (Hassig 1995; McCumber and Dryden 2022; Vadillo 2018). But while "the rat" might variously symbolize duplicitousness, corruption, greed, or a variety of other (mostly) negative qualities in various contexts, an individual rat lives out its life mostly independent of whatever such associations we might have with the broader category. Nonetheless, the material lives of rats and their cultural meanings are linked and mutually reinforcing.

That is because, through our interactions with cultural objects, we are constantly defining and redefining the material and symbolic relationships we have with them. In the course of this dance, objects may become more or less suited to the task of carrying whatever meaning we may have assigned to them. Put another way, the material qualities of an object contribute to shaping the degree to which it is "resonant" (Kubal 1998; McDonnell 2014; McDonnell, Bail, and Tavory 2017; Mohr et al. 2020) with a particular cultural use to which we might put it. This principle is demonstrated clearly by iconic art objects that require painstaking effort, specialized skill, and innovative technologies used by gallery staff to preserve their material form and its alignment with both the artist's intention and audiences' expectations (Domínguez Rubio 2014, 2016, 2020). As with fading paint on a canvas, animals are "unruly" in their materiality. A dog that unexpectedly bites its owner or attacks another dog compromises its resonance as a domestic pet. On the other hand, a rat encountered in a cage lined with wood shavings instead of an alleyway with garbage dumpsters has been removed from the category of dirty pest and is instead acting as a token for a different cultural type. This capacity highlights how cultural objects like animals are not passive receptors of shared significance, like blank canvases onto which we paint symbolic meaning, but they "act back" (Jerolmack and Tavory 2014) in ways that shape the possible contours of meaning systems. In the context of the ever-evolving relationship between materiality and symbolic meaning, then, rat extermination is both a material confrontation and a cultural narrative about rats and about our human relationships to them. Beyond simply affirming rats as killable, though, rat control efforts also generate meaning that is specific to the social and cultural context where they occur.

Taking this approach of cultural meaning to animals, and specifically to our actively antagonistic relationships with them, reveals how our dealings with nonhumans resonate not only with specific symbolic attributes but also with social structure and inequalities more broadly. Systematic extermination of another species is an expression of power and value. It is a meaning-making process that produces and clarifies a hierarchy of living things, defining some as necessitating lethal force. Occasionally, the discursive connections between this hierarchy and the power inequalities of human society are explicit and undeniable. For instance, countless organizations and state actors have produced propaganda materials that portray ethnic and racial others as anthropomorphic rats. A poster produced by the United States Information Service during World War II depicted a rat with a racist Japanese caricature for a face caught in a rat trap labeled "Material Conservation," implying that conserving resources would aid the war effort. To drive the degrading message home, the poster is titled "Jap Trap" (1941–1945). Across the Atlantic, the German Third Reich produced what are likely the most infamous examples of this genre. These included the 1940 propaganda film *The Eternal Jew*, which juxtaposed swarms of rats with the teeming populations of cramped Jewish ghettos while featuring a voiceover that directly likened the perceived threats of Jews to the disease-carrying capacity of the rodents[4] (United States Holocaust Memorial Museum n.d.). These rhetorical moves epitomize what David Pellow terms the "social discourse of animality," or the use of "nonhuman references and analogies" to convey notions of "acceptable 'human' versus nonhuman behavior and how different bodies are valued" (Pellow 2017). In other words, for propaganda like this to effectively dehumanize target demographics, it requires the premise that human life has an inherently higher value than nonhuman life, and moreover that being compared to rats is especially degrading. While there are many sources that contribute to rats' symbolic associations, the material practice of extermination crystallizes the notion that they are killable, which is in turn a designation with grave significance when it comes with symbolic analogues in human society. Not only did Hitler's Germany use references to rats to gin up antisemitism, but they would also ultimately repurpose Zyklon B, a chemical initially

---

4   Years after the dehumanizing propaganda of the Nazis, Art Spiegelman effectively drew upon and subverted these animal comparisons in his classic graphic novel *Maus* (Spiegelman 2011), which tells the story of his father's experience as a Polish Jew and Holocaust survivor. The book depicts Jewish characters as anthropomorphized mice, while Germans are rendered as cats.

developed for pest control use on insects and rodents, to murder over a million prisoners in extermination camps (Weindling 1994).

The above examples are, of course, outliers; rat imagery is not leveraged to dehumanize other humans alongside every instance of rat extermination, and it is relatively uncommon for tools designed specifically for animal pest control to be reappropriated for mass murder. Accordingly, the relationship between interspecies violence and social inequalities is not a causal, deterministic one. Nonetheless, these examples show how cultural meaning-making can bind human interactions with animals to human social structure. The power structures that shape our interspecies relationships and our social ones can be described as sharing what Bourdieu (1986) terms a "structural homology." Bourdieu uses this concept to describe the affinities between cultural "fields" that account for the correspondence between material class inequalities and analogous phenomena like tastes in food, music, or cinema. Rather than a direct, causal link between, say, income levels and preferences for popular or classical music, these fields are organized in structurally resonant and mutually reinforcing ways. Similarly, interspecies killing (especially when institutionalized and collectively accomplished) reinforces a symbolic hierarchy of species that shares a structural logic with social inequalities.

It is important to emphasize that this argument does not suggest anything approaching a moral equivalency between rat extermination and racist violence. As this book will touch on at various points, there are certainly moral questions associated with animal extermination in any given context, and some people object to all killing of animals on these ethical grounds.[5] However, these debates have defensible positions in favor of killing rats or other nonhuman animals to stop the spread of disease, protect the property and livelihood of farmers, or save endangered species, among other possible motivations. But the meaning-making effects of extermination, which generate and reinforce the symbolic hierarchy of nonhuman life, are nonetheless

---

5 Some animal ethicists even go so far as arguing that the suffering caused by interspecies predation is ethically objectionable and, if it were possible, humans would have a moral imperative to stop it. See McMahan 2010

relevant context for understanding human social inequalities. The ways we interact with nonhumans may not inevitably mirror or predict how we treat other human beings, but, as the grave examples discussed above attest, they produce a symbolic logic that can be accessed in other cultural contexts.

RESEARCH DESIGN

This book empirically investigates three separate programs of rat control, eradication, and extermination in three very different geographic locales: Alberta, Canada; Los Angeles, California; and Ecuador's Galápagos Islands. I conducted ethnographic field visits to each of these places and interviewed individuals connected to their respective campaigns against rats. I also collected archival materials and various textual data and combine qualitative analysis with computational text analysis and other quantitative methodologies. The multi-sited design of this research is intended to capture the throughlines and variations that characterize a common, widespread social practice (rat control) as observed in very different contexts.

Most commonly, ethnographic research produces in-depth, interpretive accounts of how meaning is produced within communities, social groups, or social phenomena in a single geographic locale. Within the universe of sociological methods that includes less intimate but far more wide-reaching approaches like national surveys, this is typically seen as a trade-off where generalizability is sacrificed for interpretive richness. Increasingly, however, researchers are finding this conceptualization of ethnography to be inadequate for describing the complexities of an interconnected, globalized social world (Burawoy et al. 2000). Multi-sited ethnography offers a qualitatively different approach to ethnographic research, one that attends to both the specificity of individual cases and the generality of social processes that transcend them. As Carney (2017) describes it, the advantage of multi-sited ethnography "lies in the ability to examine the interplay between micro and macro social processes." A prime example of how this proves uniquely useful for examining human-environment relationships in an age of globalized ecological crisis is Summer Gray's (2023) sprawling ethnography centered on seawalls.

Multi-sited ethnography is a particularly good fit for this topic because rat control proceeds for a variety of different reasons and with a variety of different practical implementations. Exploring these different contexts allows me to examine how rat control is at once a common, widespread phenomenon indicative of widely salient cultural meaning surrounding human-animal relationships and also a practice wherein communities and actors make meaning on a smaller, more local level. Importantly, this affords a sensitivity not only to divergent cultural contexts, but a geographic sensitivity to space and place as well. To that end, the three sites analyzed are selected to represent a typology of different landscapes where rat control takes place: rural, urban, and island landscapes.

These different landscapes shape both the possibilities for rat control and the motivations for undertaking it. In rural landscapes, rat control is motivated primarily by economic calculus. Rats themselves are imagined as threats to agricultural revenue, as pests that contaminate crop yields, sap farmers' energy, and leech off their supplies. In this context, rat control is an ongoing practice, but low population density makes maintaining large swathes of (at least mostly) rat-free territory possible. In urban areas, on the other hand, few engaged in rat control have any illusions of completely eradicating the rodents and cleansing their cities of them entirely. Instead, the goal of rat control is to protect public health by managing the spread of infectious diseases that rats transmit. This means containing rats to specific urban geographies and guarding the ultimately porous threshold between the interior and exterior of buildings. Finally, in island locations, rats are threats to cultural ideas of nature itself, as they prey on native species or damage their habitats. Conservationists work to remove rats from island ecosystems where they have been introduced by humans in the course of travel. The relative isolation of many islands contributes to motivating this goal, as islands represent self-contained landscapes that resonate with widespread understandings of nature as fundamentally external to the world of human affairs. This same isolation also makes islands unique compared to the other contexts of rat control in that they present the rare opportunity to completely eradicate a population of rats and prevent their reintroduction.

The specific sites that serve as case studies are described in detail below. While the foundation of this book is the multi-sited ethnography built on in-depth interviews and observational ride-alongs with practitioners involved in each rat control project, I supplemented these data with various other sources to get a fuller picture of the relevant cultural meaning surrounding each locale. This meant deploying several different methodological approaches, ranging from archival historical methods to quantitative analysis of existing survey data to computational text analysis based on the tools of Natural Language Processing. I will provide more detailed methodological accounting for these supplemental approaches during the chapters where they are used.

Alberta

In Alberta, my participants were government officials for the provincial rat control program. Alberta, a province in western Canada, is split between the Canadian Rockies in the west, boreal forest in the north, and prairie in the south and east. Beginning in the 1950s, the province began a campaign to remove rats from the province and prevent them from returning. This involved stopping rats' westward advancement across the prairie toward population centers like Calgary, the province's largest city. That government program continues to operate, conducting twice-yearly inspections of homes and farms along the border with Saskatchewan, as well as operating a province-wide hotline (310-RATS) that residents can call to report suspected sightings of rats.

The maintenance of a rat-free Alberta is more than just the protection of an eccentric and rather arbitrary claim to fame. In the early years of the rat program, the government launched an information campaign that included propaganda posters depicting rats as an invading army encroaching on Alberta and imploring the populace to contribute to killing them. This campaign was necessary in part because most residents in the southeast of the province, the battleground area in the campaign against rats, had never actually encountered a rat and therefore struggled to distinguish them from other, similar rodents (a phenomenon which persists to this day and continues to pose problems for Alberta's rat control personnel). Despite this difficulty of identifying the animals, the program has not just succeeded in physically keeping rats at bay, but it has also made them into a symbolic villain in Albertan cultural life and given the province a collective point of pride that is part of its regional identity. Moreover, the cultural identity of the rat as a nefarious outsider resonates with a broader cultural current of oppositional identification in Alberta, a province that has long had a contentious relationship with the rest of Canada and strives to assert its own independence and distinction.

To examine the role that this government rat control program plays in Alberta's cultural identity, I conducted an ethnographic ride-along during its inspections along the Saskatchewan border and interviewed the regional "pest control officers" who conduct them. I also conducted interviews with Alberta's provincial rat and pest specialist, the head of this program, as well as a convenience sample of Calgary residents in order to gauge contemporary public awareness of and opinions regarding the rat control effort. To supplement these ethnographic and interview data, I collected archival historical materials related to the program's early years from the Provincial Archives of Alberta, located in the capital city of Edmonton. Some of these materials included photographs, newspaper clippings, and yearly reports compiled by the province's Department of Agriculture, which housed the program in its early years. Others came from the Provincial Archives' Premier Papers collection, which contains historical materials related to the various executives that have led Alberta during its history. I collected materials including public addresses and written correspondence by Ernest Manning, the premier of Alberta during the early days of the rat program. This provided a more vivid account of the cultural and social context in the province when the program began.

Los Angeles

In Los Angeles, I interviewed a variety of city employees and contractors involved in the official response to a highly publicized 2019 rat infestation of the Civic Center, the large complex of office buildings that is home to LA City Hall. This infestation, which was reported in the *Los Angeles Times*, coincided with an outbreak of typhus in downtown Los Angeles and created a public health crisis for the city and public relations problems for the local government. One city employee who contracted typhus sued the city, claiming that their infection was the result of dangerous working conditions posed by the rat-infested Civic Center.

Months after the initial coverage of the infestation, the *Los Angeles Times* ran a follow-up story that reported on the previously undisclosed assessment of a pest control company contracted by the city to inspect the area. The company's findings explicitly linked the rat problem to homeless encampments surrounding the office buildings. Specifically, they claimed that these encampments created conditions that invited rat infestation, which included discarded food scraps and human waste on sidewalks and in planters (Smith and Zahniser 2019). With this connection, rats took on a symbolic importance in city affairs by embodying several contentious and related public issues, including the spread of infectious disease and the LA area's long-standing and fraught struggles with housing security and homelessness. Homeless encampments of the sort more common in the nearby Skid Row area had become an increasing presence around the Civic Center, and when the reports of infestation began, the "rat problem" became for all intents and purposes inseparable from the contentious politics of homelessness policy. The city instituted several measures to address the rat infestation, many of which necessarily bled over into the policy domain of homelessness. All attempted to protect public confidence in the boundary between "indoors" and "outdoors" that both rats and homelessness transgress in parallel ways.

Compared to my other sites, the problem of rat control in downtown LA is uniquely difficult to contain in a discrete organizational project, and instead overlaps with myriad social problems and areas of bureaucratic jurisdiction. While the city maintains a contract with a private pest control company that conducts inspections around the Civic Center, keeping rats at bay necessarily involves other organizational bodies, including the Departments of General Services (GSD) and Personnel, the Department of Sanitation, Parks and Recreation, and the LA Police Department, among others. I interviewed employees from GSD, Sanitation, the office of an LA city councillor, and the private pest control company contracted by the city. I also conducted participant observation with the Sanitation Department, which conducts weekly cleanings of the Civic Center that were instituted in response to the rat infestation. Finally, to empirically analyze public discourse on LA's homelessness issue, I collected a sample of over 1,000 articles in the *Los Angeles Times* on this topic, which I examine using a mixed-methods content analysis.

The Galápagos Islands

In the Galápagos Islands, rats are just one of the species targeted for eradication by conservation initiatives hoping to protect native species and their habitats. In the 1990s and early 2000s, it was goats that occupied most of conservation groups' energies, as a collaboration of the Galápagos National Park and various nongovernmental organizations attempted to completely remove them from the island of Isabela. This project was controversial for the extreme methods it employed, which involved sharpshooters exterminating the animals with high-powered rifles and hormone treatments designed to make the goats more social and thus easier to dispatch with in higher numbers at once. Such methods exemplify the lengths to which conservationists will go to return the island landscape to a particular physical and ecological state. The eradication of rats that brought me to the Galápagos is extreme in another way: it is the first program on an island with the complicating factor of a significant permanent human population.

While these programs are invested in protecting nature, they simultaneously contribute to defining both ideas and experiences of nature itself. The Galápagos Islands are iconic for the wildlife they are home to, in no small part because of their mythologized role in inspiring Charles Darwin's theory of evolution. As I will examine in detail, native species of the Galápagos, especially the Galápagos tortoise, symbolize nature more thoroughly than perhaps any other type of wildlife because they have become synonymous with the process of evolution, emblems of the very laws of nature themselves. This introduces a contradiction that frames modern conservation attempts: nature is iconic for representing processes of change, but protecting nature often amounts to preventing change. Put another way, conservationists hope to undo the "unnatural" influence of humans and human-introduced species like rats in order to return the landscape to a preferable state. This begs the question of who or what can be a legitimate agent of environmental change, and what is the appropriate role of humans in revered natural landscapes.

To investigate these tensions, I interviewed employees of the NGO Island Conservation, which partners with the Galápagos National Park and other conservation groups on species eradication programs for conservation goals. Interviewees included conservation biologists, wildlife ecologists, environmental lawyers, and project managers, among others. I traveled to Floreana Island to visit the site of the current eradication program, where team members were preparing for "implementation," or the beginning of the poisoning program, which was then two years away.[6] A major part of this preparation is the construction of chicken coops and other structures to house livestock during the poisoning program, as well as aviaries, which will house entire populations of native bird species to ensure that they are not unintentionally poisoned. I toured these construction projects and spoke with the local population working on them. In addition to this field visit and the interviews conducted in the Galápagos Islands, I also interviewed an Island Conservation employee based in the United States who helps coordinate Project Floreana from afar, as well as a conservation biologist and professor emeritus who pioneered methodologies employed by the organization. In addition to these interview and ethnographic data, I also analyze planning documents obtained from the Galápagos Conservancy, one of the organizations involved in these programs, which pertain to specific eradication projects. These data offer important insights, as they represent direct articulations of the programs' goals, methodologies, and justifications in the practitioners' own words. Finally, I also collected a sample of over 1,000 blog posts from the websites

---

6   As with many things scheduled for 2020, the implementation of the baiting and poisoning program was delayed due to the COVID-19 pandemic. Implementation ultimately began in October 2023.

of the various conservation organizations involved in the eradication. I use a combination of Natural Language Processing techniques and qualitative coding to examine the text of these blogs and produce a picture of how conservationist organizations imagine the natural ecosystems they seek to protect.

PLAN OF THE BOOK

The rest of the book is divided into three parts, each comprising two chapters. The three parts correspond to the three research sites that serve as case studies of rat extermination. In addition to functioning as self-contained empirical case studies, each part advances a theoretical insight that contributes to the book's overarching arguments.

Part I, "Animals in Boundary Work," explores the case of Alberta's long-standing rat control program and considers these questions: Why is being "rat-free" so important in Alberta, and what does "the rat" mean to Albertans as a cultural object? In the process, these two chapters advance a theory of animals' role in "boundary work" or the social negotiation of group identities and their limits. I argue that, as cultural objects, animals participate in boundary work in unique ways owing to their mobility and sentience. These qualities lend a spatial component to our symbolic relationships to them that in turn contributes to inscribing the boundaries of group membership on the physical landscape.

Chapter 2, "Rats and 'Boundary Work' on the Canadian Prairie," introduces Alberta and its rat control program, delving into the unique and often antagonistic relationship the province has to Canada at large. This relationship reveals how Alberta has historically defined its collective identity in an oppositional manner, asserting its independence and distinctiveness from the rest of the country. In this context, the province's rat control program performs salient cultural meaning-making that resonates with prevailing stories of self-identity in Alberta. I analyze ethnographic data from participant observation with pest control officers on the Alberta prairie, showing how rat control gives salience to Alberta's physical geographic borders by adding a distinguishing element (the quality of being "rat-free") to an otherwise indistinguishable landscape. Far from an arbitrary claim to fame, though, Alberta's "rat-free" status is a narrative of moral accomplishment tied to the cultural meaning of the rat itself as a symbolic villain, as evidenced by archival data. Human-animal interactions are uniquely important to the negotiation of group identity and its boundaries because, as both cultural objects and elements of the broader biophysical landscape, our relationships with animals are both spatial and symbolic in nature. As such, paying attention to the role of human-animal relationships provides important insight into how group identity and the limits of group membership are achieved through an interrelated negotiation of spatial, geographic borders and symbolic, discursive ones.

Chapter 3, "'You Can't Ignore the Rat': Guarding Alberta's Moral Character," delves more deeply into the notions of moral purity that have historically characterized Albertan collective identification and how the province's rat control program resonates with them. I compare the construction of the rat as an external menace that must be controlled to potent currents of anti-immigrant sentiments in Alberta. The pride that Alberta takes in its rat-free status begs the question of what else Alberta's notion of collective identification is invested in keeping out. The early years of the rat control program featured an awareness campaign that not only informed Albertans of their government's priority of making the province rat-free, but also sought to create enthusiasm for this effort by presenting rats as a nefarious invading force looming on Alberta's borders. This media campaign resonated with various incarnations of nativist politics in the province. Alongside these posters and other rat control–related materials from the 1950s, chapter 3 analyzes other archival materials that shed light on the cultural and political climate in Alberta that provided the context for the early years of the program. The provincial government that oversaw the initiation of this campaign, these materials show, was deeply concerned with other perceived threats to Alberta's moral character, both internal and external. Chief among these was communism, which Alberta's premier, Ernest Manning, regarded as both an insidious, godless force that might compromise the Christian values he aspired to for Alberta and also an external menace in the form of a nuclear USSR. This same period coincided with some of the most prolific years of Alberta's long-standing forced sterilization program, which disproportionately affected certain minority groups. In this historical context, analysis of the rat control program reveals how, as a cultural narrative, it is premised on a moral hierarchy of species that mirrors and mutually reinforces the social power inequalities of the province.

Part II, "The Borders of Urban Nature," further explores how rat control operates in the negotiation of boundaries by highlighting the underlying importance of one particularly salient distinction: the border between nature and society. Through analysis of a 2019 rat infestation in Los Angeles's Civic Center complex, rat control emerges as crucial for guarding a key spatial logic that guides the management of urban nature: the boundary between "indoors" and "outdoors."

Chapter 4, "Rats and the Indoors/Outdoors Divide," introduces the case of Los Angeles's City Hall rat infestation, the city's response to it, and the broader cultural context of LA's ambivalent relationship to nature. When City Hall was reported to have a rat problem, the news came amid an outbreak of typhus in downtown LA. Chapter 4 draws on interview data with city officials and secondary literature to theorize this issue as a symbolic and material negotiation of urban nature. Whereas "good" forms of nature have become central to urban imaginaries, crafting a desirable, ideal urban environment means keeping "bad" nature like rats at bay also. This, I argue, is accomplished largely by protecting public faith in the threshold between "indoors" and "outdoors," which functions as a crystallization of the broader nature/society dualism. Ideally, cities must bring good nature to the outdoor areas of cities while sufficiently keeping bad nature, like disease-spreading rodents, from breaching the indoors. Rat control, therefore, is an effort not just to protect city officials from infectious diseases, but to clarify and enforce a moralized categorization of the nonhuman world.

In chapter 5, "Bulky Items: The 'Rat Problem' and the 'Homeless Problem' in Downtown LA," I examine how the public issue of this rat infestation bled into other arenas of the contentious politics of urban space in LA. Specifically, in the months after the infestation, the *LA Times* reported that, in its official report on the issue, the private pest control company contracted by the city directly tied the rat infestation to these homeless encampments. Much of the city's response to the rat issue in the Civic Center has either implicitly or explicitly involved the homeless, and discussions of the "rat problem" in my interviews often veered toward the topic of the "homeless problem" to the point where the two became nearly inseparable and at times even indistinguishable. Whereas rats entering City Hall represent an unacceptable breach of the "inside" by the "outside," the homeless encampments around the Civic Center, and, importantly, the storage of "bulky" personal belongings in public space, represent an inverse of that transgression, where elements of the "inside" inappropriately reside "outside." The reciprocal spatial transgressions of rats and homelessness come to be symbolically tethered to each other as inseparable public issues. Moreover, through discourses of public health (concern over rat-borne diseases like typhus, and later over COVID-19), the spatial logic that is maintained through the interconnected projects of rat control and homeless policy becomes a moral prerogative. Drawing on a mixed-methods content analysis of *Los Angeles Times* articles and an ethnographic ride-along with LA sanitation workers, chapter 5 examines the spatial management of homelessness and the spatial management of rats in LA. This analysis reveals how, in practice, managing the indoors/outdoors divide requires enforcing the unequal politics of access to public and private space.

Part III, "Ecologies of Meaning," considers the negotiation of the nature/society boundary at the broadest level by examining the case study of species eradication programs for environmental conservation. These two chapters make sense of the moral contradictions that arise in environmental conservation in terms of what I refer to as "ecologies of meaning," or the system of interconnected cultural objects that make up broader holistic cultural ideas of "nature."

Chapter 6, "Ecologies of Meaning in Ecuador's Galápagos Islands," explores the current rat eradication efforts on Floreana Island in the context of the human history of the Galápagos Islands. The cultural significance of conservation work is heightened in the Galápagos compared to other locales because of their enduring associations with Charles Darwin and the theory of evolution. Because of the potent, mythological importance of Darwin's legacy on the

Galápagos, the islands' native species have become living cultural icons of the laws of nature themselves. In terms of material culture, conservation programs attempt to protect or reestablish the material conditions under which these animals can retain the resonance needed to function in this symbolic role. Somewhat paradoxically, this often means seeking to restore a prior version of the ecosystem, though change itself is central to the theory of evolution that makes the nature of the Galápagos so iconic. I argue that conservationists must negotiate good and bad nature by delineating between acceptable and unacceptable environmental change. In doing so, they intervene not just in the material ecology of the island, but also in the symbolic landscape that overlays it, the ecology of meaning.

Chapter 7, "Killing for Life: Morally Acceptable Lives and Deaths in Environmental Conservation," examines how the cultural meaning that guides this conservation work is morally inflected and species eradication is necessarily a question of life or death. When is lethal force, whether by rifle or by poison, an appropriate tool in environmental stewardship? Chapter 7 focuses on how conservationists grapple with this moral question, and how they have responded to detractors, especially those concerned with animal rights. Ultimately, this reveals the contested nature of the ecology of meaning, where different regimes of value must negotiate the symbolic meaning of animals and ecosystems, and how, in negotiating these inherently contradictory questions, conservationists implicitly tread into broader ethical terrain around what bodies (animal or otherwise) are dispensable or worth protecting. At its core, this is a negotiation of the boundaries between nature and society, or more specifically between good and bad nature. Chapter 7 concludes by considering possible technological innovations in species eradication based on genetic editing techniques, which promise an end run around the question of violence. The core question of the boundaries between good and bad nature, however, remains. This underscores how power inequalities will not only determine whether humanity can meet the challenges of climate change and our age of ecological crisis, but will also shape the possible environmental futures that will emerge by doing so.

The book concludes by revisiting the three cases of rat control and the theoretical takeaways of each. The conclusion elaborates how each case contributes to the overarching arguments that (1) rat control negotiates the boundaries between nature and society; and (2) this process generates interconnected hierarchies within and between species. I close by considering the implications of these arguments for the future of our relationship with the nonhuman world.

PART I

ANIMALS IN BOUNDARY WORK

× TWO ×

RATS AND "BOUNDARY WORK" ON THE CANADIAN PRAIRIE

"March is typically a miserable month" on this rural expanse of Canadian prairie, according to Jesse, a pest control officer (PCO) for the area. It coincides with the toilsome labor associated with calving season for the many cattle farmers in the region, and temperatures in Alberta reliably remain frigid this time of year. Sure enough, spring was late in arriving on one cold March morning I spent in Eastern Alberta, roughly twenty miles from the Saskatchewan border. I had arrived there the previous night with Phil Merrill, Alberta's provincial rat and pest specialist, after driving roughly three hours from Calgary. Though the journey was a straight shot across a highway that seemed like its route might have been drawn on the map with a ruler, the conditions made it harrowing at times; icy roads and snow flurries had left one of the few other motorists we encountered marooned in a snowbank. Thankfully, Merrill's sturdy pickup truck delivered us safely to our destination, a small farming town whose population may have doubled that night, as it was hosting a regional youth hockey tournament. That day we were to meet Jesse, who was in charge of ensuring that the strip of land here along the border is free of rats. When I first heard of Alberta's claim to "rat-free" status, I had the same reaction of incredulity that so many others have when learning of it. The idea that such an enormous swathe of land could be free of rats despite having a human population of over four million people, one that included two major metropolitan areas, no less, was baffling. But when I looked out onto the frozen prairie, where it was difficult to tell where snow ended and overcast sky began, it was hard for a Southern Californian like me to imagine the environment being hospitable to much at all, even the famously hardy and adaptive rodents.

Alberta is over 255,000 square miles, a little over 1.5 times the area of California, but the region I was visiting is by far the biggest area of concern for the rat control team. Occasionally, Merrill tells me, he and his team must deal with a single rat that hitches a ride on a truck or an RV into Calgary or another population center in the middle of the province, but those solitary rats seldom, if ever, turn into infestations. The cold, alpine conditions in the Canadian Rockies along Alberta's western border with British Columbia, and the sparsely populated and well-managed prairie of Montana to the south, are poor conditions for rats. The same is true of the frigid far north of the province where it gives way to Canada's Northwest Territories. It is this troublesome eastern border, the straight, invisible line through the prairie that separates Alberta from Saskatchewan, which occupies most of Merrill's energies. On both sides of this line there are grain and cattle farmers whose properties may contain the warm environment and sources of food that rat populations require to get established. Should that happen on the Saskatchewan side, it takes the vigilance of Merrill and his team to ensure that the animals do not spread into neighboring farms across the border. Each Municipal District in the area along the border, deemed the "Rat Control Zone" (RCZ), has a designated PCO in charge of conducting these inspections and taking preventive measures against infestations.

As I explore in this chapter, Alberta's rat control program not only monitors the province's eastern border for rats, but also, in doing so, maintains several other meaningful boundaries in Albertan cultural life. By enforcing a set of standards around the practice of agriculture, it most fundamentally maintains the terms of an acceptable relationship between human society and the natural environment. In doing so, though, the program also becomes crucial to a set of related spatial and discursive boundaries of group identity. Specifically, the regular work of inspecting the border with Saskatchewan for rats gives salience to these physical, geographic borders of the province by adding a distinguishing element (the quality of being "rat-free") to an otherwise indistinguishable landscape. Far from an arbitrary claim to fame, though, Alberta's "rat-free" status is a narrative of moral accomplishment tied to the cultural meaning of the rat itself as a symbolic villain. This makes rat control a particularly resonant practice in Alberta, where collective identification has historically been potently oppositional in character.

In a broader sense, I argue that nonhuman animals in general are uniquely suited to participating in the type of boundary work I describe above, because they are simultaneously cultural objects and sentient elements of the biophysical landscape. Because of this, our relationships with them are both spatial and symbolic. The rest of the chapter explores this notion primarily through ethnographic fieldwork with Alberta's rat control program. I first provide overviews of both the concept of boundary work and animals' unique place in it, as well as Alberta's history of oppositional identification, characterized by an antagonistic posture toward the rest of Canada that the province has frequently adopted.

RAT CONTROL AND "BOUNDARY WORK"

Alberta's rat control program is part of a narrative process of meaning-making in which Albertans define their collective provincial identity. Put more simply, who we are is (in part) the stories that we tell about ourselves, and the rat control program is one such story, or rather, a collection of stories. "Alberta is rat-free" and "rats are menaces to society that must be killed," for instance, are narratives that contribute to establishing a sense of collective identity and shared values. As I will discuss, Alberta's "rat-free" status is not absolute, but rather is more fluid than the phrase would suggest. Nonetheless, as a narrative, the notion of a rat-free province is a meaningful and powerful claim in the cultural imagination, even if its material reality is more complex.

More broadly, this reveals how animals shape the negotiation of group identity through "boundary work," or the mechanisms by which people imagine and negotiate the limits of social categories and intuitively understand the lines between "us" and "them" (Lamont 1995; Tilly 2004). Group categories (in terms of such axes of identity as ethnicity, religion, or class status, for instance) are dynamic and constantly renegotiated through collective meaning-making. For instance, sociological scholarship has taken up how sonic styles (Schwarz 2015) and aesthetic judgment (Wohl 2015) can be central factors in consolidating group identity, and how boundary work itself might be delegated to "anonymous others" (Tavory 2010). I argue that animals participate in boundary work in unique ways because, as both cultural objects and elements of the broader biophysical landscape, our relationships with them are both spatial and symbolic in nature. As such, paying attention to the role of human-animal relationships provides important insight into how group identity and the limits of group membership are achieved through an interrelated negotiation of spatial, geographic borders and symbolic, discursive ones.

It could rightly be pointed out that other elements of the biophysical landscape, like rivers, trees, mountains, or even buildings, also certainly function as symbolic objects, and we do of course relate to all these both spatially and symbolically. Nonetheless, animals are unique among this list for their sentience and for their active capacity in social life. As totemic symbolic objects, they are elevated to the level of near-human, quasi-anthropomorphized characters. Whereas the oak tree surely symbolizes sturdiness, the rat on the Canadian prairie acts out its role as a demonized villain with sentience (however willingly or unknowingly) when it crosses the Saskatchewan border and vexes Albertan farms.

It is this combination of mobility and symbolic potency that makes animals salient figures in the related processes of boundary work and placemaking. Geographer Doreen Massey (1994) influentially argued for a retheorization of place as increasingly formed through movement rather than fixity, anticipating a subsequent boom in scholarship from other human geographers on "mobilities" (Cresswell 2006; McCumber 2018; Sheller 2016; Sheller and Urry 2006; Urry 2007). Massey's argument diverges from those of scholars like David Harvey, that accelerated globalization has rendered placemaking an exclusionary and reactionary process wherein people draw boundaries to assert a fixed, "militant particularism" in the face of dizzying and homogenizing movement and change (Harvey 1996). While Harvey's notion, borrowed from Raymond Williams, appears increasingly prescient in light of numerous instances of reactionary placemaking projects built around borders (nativist calls for a US-Mexico border wall being a case in point), the negotiation of these physical boundaries and the collective identities they symbolically enclose is borne out through the movement that Massey emphasizes. Animals' mobile symbolism in particular and our human interactions with their movements are a key social process in drawing these lines of demarcation.

Specifically, in Alberta, rat control consolidates notions of "Albertan" collective identity by reinforcing the spatial relationships that make Alberta's actual provincial boundaries meaningful distinctions. It is the totemic symbolism of rats, though, that makes these borders mark more than just an arbitrary distinction. That is to say, the meaning of rats themselves gives meaning to the idea of a rat-free province. By drawing a line in the prairie and systematically defending it against rats, Alberta's program both continuously constructs the rat as a symbolic villain worthy of lethal force and, in the process, maintains a cultural narrative that doubles as an identity-defining moral claim ("Alberta is rat-free"). The relationship between rats and Albertan collective identification recalls Tim Cresswell's (2014) argument, drawing on Mary Douglas, that a meaningful sense of place often coalesces most potently around notions of what lies "outside" or out of place. Moreover, the interplay between meaning and materiality involved here highlights animals' spatial and symbolic capacities; the symbolic identity of the rat transcends its physical, material reality, but the narrative of a rat-free Alberta is deeply tied to both rats' spatial distribution *and* their cultural meaning.

HISTORICAL BACKGROUND: OPPOSITIONAL IDENTIFICATION IN ALBERTA

Placemaking and collective identification in Alberta must be understood in terms of the province's uneasy relationship with Canada at large. The provincial government of Alberta has a long history of clashes with the Canadian federal government and has frequently attempted to assert its own autonomy and resist what Albertans feel are undue impositions by the Canadian state. A pivotal moment in the history of these tensions was the enactment of the National En ergy Program (NEP) by Pierre Trudeau's Liberal government of the 1980s, which "included policies to regulate the prices of oil and gas in Canada, to impose several new taxes on the oil and gas industry, and generally increase the federal role in the sector" (Cairns 1992). The legislation was severely unpopular in western Canada, and particularly for Albertans, a large portion of whom saw it as exploiting their province's robust energy sector on behalf of eastern provinces and at the expense of Alberta's own interests.

The resentment of the NEP contributed to a political and cultural discourse in Alberta and neighboring provinces known as "western alienation," which Roger Gibbins has described as "a political ideology of regional discontent . . . [encompassing] a sense of political, economic, and . . . cultural estrangement from the Canadian heartland" (1980: 169). Gibbins goes so far as describing western alienation as a "cloak or costume" put on by Albertans "to position themselves with respect to the national community; they learn its tenets as they emerge from the egg, and seldom lose faith no matter where in Canada they might come to live" (Gibbins 1992: 70). While Albertan culture is not monolithic, the ideology of western alienation and a related sense of resentment for the mandate of Canada's national government are foundational to the conservatism that enjoys preeminence in the province's political climate.

In terms of national politics, this hostility between Alberta and the Canadian federal government has manifested in various ways. At times, Alberta and its neighbors projected a desire for a greater voice in national affairs. For instance, the 1980s saw the emergence of a populist right-wing political party called the Reform Party, which used the slogan "The West Wants In" for its insurgent shake-up of Canadian national politics. At other times, though, the current of western alienation has been channeled into movements for autonomy for Alberta and related assertions of its distinct cultural identity. A political movement called "Alberta Agenda," founded in 2000 by a group of Alberta conservatives that included future prime minister Stephen Harper, is a key example of this desire for autonomy in the province. The movement's political message is perhaps most neatly summarized by the text of a prominent billboard on Alberta's Highway 2, which I passed countless times during my fieldwork and was my first introduction to Alberta Agenda. The billboard simply read "MORE ALBERTA LESS OTTAWA." Alberta Agenda introduced its policy aims in a letter known as the "Firewall Letter." The document's name comes from a passage that reads: "It is imperative to take the initiative, to build firewalls around Alberta, to limit the extent to which an aggressive and hostile federal government can encroach upon legitimate provincial jurisdiction."[1]

The cultural sentiments of western alienation are certainly not confined to Alberta alone, but as Alberta Agenda's rhetoric demonstrates, they are a particularly strong current in Alberta's provincial politics (and are enflamed by the economic tensions surrounding the province's fossil fuel industry and Canada's national energy policy). The notion of its unique and alienated position in the country extends beyond the arena of politics, as prevailing cultural narratives in Alberta have consistently cast it as "a maverick agrarian region that is distinct ... from the rest of Canada" (Blue 2008: 74) with its own unique iconography. While popular imaginations of Canada that proliferate in the United States may cast the country as culturally unified under the iconography of ice hockey, the maple leaf, and Tim Horton's donuts, Steven Penfold writes that Canadian "ideas of national identity ... are refracted through regional attachments" (2002: 26). In Alberta specifically, both the character of its collective identification and the discursive mechanisms that construct it are oppositional in nature; Alberta is constructed in the regional cultural imagination by a meaning-making process that explicitly sets the province apart and rejects the eastern, Ontario-centered hegemony. For instance, Blue (2008) examines how "Alberta Beef" has been elevated from an otherwise arbitrary agricultural commodity to a salient marker of Alberta's regional identification that signifies the masculine, rural cowboy culture at the heart of Alberta's cultural mythology and distinguishes it from Canada's cultural identity at large.

As will be examined in greater detail in chapter 3, Alberta's rat control program began in the 1950s, well before the NEP enflamed tensions between western Canada and the federal government, but

---

1   The text of the letter can be found here: http://westernstandard.blogs.com/shotgun/2010/03/the-firewall-letter.html.

its inception can be understood as part of the province's overarching will to act as a distinct entity. As Phil Merrill narrated it to me in one of our interviews, the success of the program benefited from some serendipitous timing in terms of both rats' progression across the Canadian Shield and the development of the province's administrative capacities. He contrasted the governmental development of Alberta during the beginning of its rat control program with the situation in neighboring Saskatchewan when rats first arrived in that province two decades prior: "Their government wasn't very organized.... They didn't have many departments. When they [rats] hit our border in the '50s, we were just barely getting going. We had a Department of Health, we had a Department of Agriculture and we could find the money to put up the program. Whereas if they had hit us in the '30s, they would have come right over. It was just kind of a timing thing."

This particular historical narrative of the program casts it as an early part of Alberta's project of consolidating its collective identity by using the organization of its provincial government to assert its geographical boundaries. The physical distribution of rats became a key factor in establishing those spatial boundaries that affirm Alberta as a distinct place with a coinciding collective identity. This is not to say that the project of rat control has been explicitly motivated by a desire to consolidate a quasi-nationalist regional identity for Alberta. The program was started by the Department of Health out of concern that rats might spread plague and was eventually turned over to the Department of Agriculture, when it was decided that the biggest threats the animals posed were to infrastructure and to farmers' economic livelihoods.

Nonetheless, I will argue in this chapter and the next that rats and rat control have functioned as a tool of identity formation in Alberta because the program, while motivated by these earnest concerns around public health and economic productivity, is particularly resonant with Alberta's broader oppositional identification rooted in boundary work. As discussed above, a potent aspect of Alberta's cultural mythology is the notion that it is a distinct geographic, political, and cultural entity. Likewise, the rat control program is a cultural narrative of regional exceptionalism that affirms a "sense of place" in Alberta by reinforcing its discursive and spatial boundaries.

This regional identification is built on insider-outsider relationships that can sometimes be nebulous; the "outsiders" in this case are most easily identified as the political and cultural elite of eastern Canada who might impose undue influence on Alberta, while, by comparison, the province shares an affinity with neighboring Saskatchewan for their similar experiences of alienation. This makes the process of boundary work more nuanced, as Alberta's project of oppositional self-identification and placemaking proceeds in a way in which its most proximate neighbors are not necessarily the most potent objects of that opposition. The rat thus functions as a useful symbolic "outsider" against which Alberta can clarify its geographical and cultural boundaries. The placemaking effectiveness of the spatial process of rat control is directly tied to the related construction of the rat as a vilified cultural object. As I will show in chapter 3, this is especially apparent when examining the shifting cultural meaning of rat control in light of emergent demographic changes and related contestations over collective identification. While these changes indicate contemporary challenges to the resonance of rat control as a cultural narrative, the program maintains a prominent place in Albertan public life.

RAT CONTROL ON THE ALBERTA PRAIRIE

Save for its western border with British Columbia in the Canadian Rockies, which is partly drawn by the Continental Divide, Alberta is defined by rectilinear boundaries. This allows the rural farmland in the east to be easily sectioned into a grid. This system predates the advent of the rat control program, but it is used to systematize rat inspections and prevention. Because the main risk of infestation comes from the border with Saskatchewan, the most important lines of demarcation are the "ranges" that correspond with east–west distance from that border. Each range is six miles wide, meaning "Range 1" refers to the narrow stretch of land within six miles of Saskatchewan.

This system was explained to me on my ride-along in the RCZ with Jesse, the PCO in charge of inspections in this area, and Phil Merrill, Alberta's provincial rat and pest specialist, as we all drove east on Highway 9. Jesse noted that we were then in Range 3, where we stopped at a few properties that were quickly determined not to have rats. At the first of these, a small house with a shed, I was a bit surprised when we did not even bother leaving the truck. From the driver's seat, Jesse pointed out that the floor of the shed was lifted a foot or so off the ground, which he said would make the temperature inside unsuitable for rats. He also noted that it was clear from this cursory visual inspection that the owners of the property are in and out of the shed frequently for work on their farmland. Given this, coupled with the fact that Jesse has a close relationship with these residents, he reasoned that if there were a rat they would almost certainly have seen it and notified him already. The next property was a granary with several bales of hay stacked near it in the middle of a field. This time we ventured out of the truck, but we mostly did so just for my education, not because Jesse had any reasonable suspicion that there might be a rat here. Jesse explained to me that he had left poisoned bait on these bales not long before, and that the snow there was fairly fresh, so if a rat were there, we would see tracks. To make it clear just how little he was concerned about this site having a rat problem, he also added, "I probably baited this when I was bored."

For Jesse and other PCOs who conduct these inspections, rats (or, more accurately, the absence of rats in common cases like these initial visits) are instrumental in establishing a social relationship with and definition of the landscape. While these inspectors are the people who most directly carry out the work of rat control, their efforts are contingent upon the cooperation and investment of the local residents, with whom the inspectors typically have close relationships. The absence of rats is mapped onto a set of individual and collaborative practices that more broadly serve to maintain the terms of an acceptable human-environment relationship. From small details under individual discretion like the height of a shed off the ground, to factors at the institutional level like the way these municipalities dispose of garbage, various practices combine to form a meaningful social process that squares the socio-environmental order here with prevailing sets of cultural values. The rat control program is a central part of this process, as the "rat-free" status is a point of collective identity and pride, stemming from the maintenance of a particular state of human-nonhuman relations. Hence, while PCOs like Jesse ensure that the area is rat-free, they also undertake a more general community enforcement of the socio-environmental standards here.

Another PCO, Dave, brought up a specific example to illustrate the spirit of this collaborative effort during an interview with him and Phil. A property in this area had an old structure that became infested with rats and had to be burned down. Bob, a resident of the area, had frequently assisted the PCOs in charge there by conducting inspections when they were busy or out of town, and Jesse contacted him for help with this job. Phil recounted in detail: "Bob brought his fire truck from the county and a crew so that when we were burning the building, if we had problems, he had the fire truck there. So it wasn't just Jesse looking after the rats. Jesse was killing the rats, but the [municipality] provided the manpower and the equipment, not only just the fire truck but they brought a pay loader to push the remaining part of the building into the pit and fill it in and move the steel granary away. So it's a cooperation."

During my ride-along, I experienced another instance of this sense of community collaboration. On one of our stops to inspect some bale stacks on a farm, Jesse's truck got stuck in a deep pocket of snow, its wheels spinning as he tried to back up onto the road. Jesse was more annoyed than concerned, and he contacted a friend and hockey teammate of his who lived nearby. After roughly fifteen minutes, Jesse's friend arrived with a large tractor to pull us out. While Jesse was not enlisting help specifically for rat control this time, the incident demonstrated how Jesse relies on the broader community in the area for assistance in his PCO duties. Beyond these occasions where he requires their direct help, he also depends on them to do the everyday upkeep and maintenance of their farmland and properties in accordance with the rat control program's guidelines.

While for outsiders the cultural narrative of a rat-free Alberta may suggest that the Alberta-Saskatchewan border is marked by an impenetrable rat-proof fortress, the residents near the border who participate in these collaborative efforts are well aware of the effort that goes into continually maintaining their province's rat-free status. In fact, Dave and Phil acknowledged the fluidity of that status in the Rat Control Zone by noting rats' mobility. Dave began by conceding that Alberta's "rat-free" status is not absolute: "You know we're here on the border.... For the amount of feed that's trucked through Alberta, coming out of Saskatchewan, and the amount of equipment that comes out of Saskatchewan into Alberta—cattle liners, you know, even personal vehicles—[a rat is] liable to catch a ride, right?"

What Dave's observation illustrates is that the notion of a "rat-free" Alberta is not a static state of affairs but rather an ongoing process that necessitates continual maintenance. In this respect, it is best understood as a cultural narrative that, while rooted in a material status, simultaneously elides the complexities of that status. Phil provided further specificity on this point: "We have two definitions of 'rat-free.' One is we're rat-free right now, until we get the call that there's a rat and then we go exterminate that rat and then we're rat-free again. Another definition of 'rat-free' means we don't have a breeding population of rats. So, at one point in time we might have an infestation that's a breeding population, but we get rid of that within a very short period of time and then we don't have a breeding population."

Rat control maintains the efficacy and symbolic importance of the broader cultural narrative of a "rat-free" Alberta. Most people may not have the subtle distinctions Phil notes here in mind when they think of Alberta's rat-free status, but even with these qualifiers, it is the maintenance of the dynamic conditions in the Rat Control Zone that makes the cultural narrative meaningful for the rest of the province. Though the material reality of the "rat-free" status is fluid, Dave was adamant that there was a palpable sense of pride around it on the Alberta side of the border: "You'd be surprised how [proud] people are—if you were to ... go straight north and south [from here], they're pretty happy about it, that they can say they're rat-free. For sure."

Nonetheless, Phil indicated that the narrative of a rat-free Alberta is most compelling for those Albertans who are least often confronted with the fluidity of that status. In other words, there is a spatial component to the narrative's salience, wherein it is stronger farther away from the Rat Control Zone: "I think the only ones that would be a little bit not so proud are these guys right along here that get the occasional infestation who say, 'Oh, well, we're not really rat-free—

I had rats five years ago.' . . . You know, 'I'm always catching a rat.' They probably don't have quite the same pride as somebody further out that says, 'Hey, well I don't [ever] have to worry about rats.'"

The fluidity of the province's rat-free status aside, the inspection work in the Rat Control Zone is still central to the symbolic weight of the cultural narrative of rat control. In fact, rather than diluting the narrative of Alberta's rat-free status, the acknowledgment of infrequent infestations on the Alberta side and of the human mobility that facilitates them actually demonstrates the symbolic weight of that narrative; while the border is porous to the point of being almost unnoticeable for human beings, it is made a meaningful boundary by the stark difference it represents in terms of the distribution of rats. Put another way, the easy movement of people and materials across the border threatens the idea of "Alberta" as a distinct place with an associated identity, but the rat control program functions as a placemaking narrative. Even through occasional infestations on the Alberta side may mute the sense of pride in being rat-free that the residents closest to Saskatchewan feel, these inspections still make the border between Alberta and Saskatchewan a stark divide with respect to the likelihood of a rat sighting.

This principle was best demonstrated to me on a brief detour into Saskatchewan during my ride-along. After a few uneventful stops in Ranges 1, 2, and 3 on the Alberta side of the border, Phil and Jesse seemed determined to find a rat while I was with them, as if my trip there would be something of a waste otherwise. The easiest way to accomplish this, they reasoned, was to simply cross the border and leave the rat-free confines of Alberta. Save for a small roadside sign welcoming us to Saskatchewan, there was little indication when we crossed the border; the sprawling prairie landscape is more or less identical on either side of the 110-degree-west longitudinal meridian that separates the two provinces. We ventured a few miles into Saskatchewan farmland, and, as we pulled up to a farm mere minutes after crossing over, Jesse received a video message from a resident of the property, a friend of his. The video showed a dog chasing a rat across the white expanse of snowy grassland before ultimately catching and killing it. After viewing it on Jesse's phone, we looked up and noticed the same dog from the video, running through the snow-covered prairie chasing yet another rat. The owner of the dog who sent the video was busy moving bales of hay with a large tractor. Nearly every time the machine sank its metal teeth into a bale of hay to lift it, this would uncover another rat, which would scurry away and be pursued by the dog.

The ease with which we were able to find a rat infestation mere miles into Saskatchewan is an indication of how rats play a key role in making the provincial border a meaningful boundary not just for the political organization of the Canadian state, but also for the cultural lives of the people who live near it. Saskatchewan and Alberta residents in this area share the same environmental conditions and have cultural, economic, and lifestyle similarities that engender mutual affinities transcending the provincial boundary. After all, we only showed up on this property in the first place because Jesse and this Saskatchewan farmer were friends. However, it is through these same aspects of commonality that rats operate to make "Albertan" versus "Saskatchewanian" distinct identities here. One of the main impediments to making this area on the Saskatchewan side reliably rat-free, according to Merrill, is that rat infestations cannot be effectively quelled by treating individual sites if neighboring properties also have rats. In fact, the infrequent infestations that happen in Alberta typically result from infestations in bordering properties on the Saskatchewan side, but when this happens, Alberta's rat control team dispatches with the problem before it spreads farther.

Jesse tells me that often people in this rural area may not freely discuss it if they have a rat problem, for fear that it will reflect poorly on them as a farmer, noting that it might be seen as "irresponsible." I heard this sentiment in various other contexts during my trips to Alberta. Merrill would later tell me that there is a "stigma" associated with having rats on your farm, and that Saskatchewan residents sometimes react defensively to Alberta's boasting about its rat-free status, sometimes even challenging the validity of such claims. Another interviewee, a professor at the University of Calgary and a lifelong resident of Alberta, tells me that she grew up near the Saskatchewan border, where people on the Alberta side would refer to their Saskatchewanian neighbors with the mildly derogatory term "rat-landers," a variation on "flatlanders." The set of values around what it means to be a responsible farmer is not particular to either side of the provincial boundary, but in Alberta there is a long-enough-established system in place, with regular inspections and cooperation between Merrill's staff and the residents, that infestations do not take hold. There is thus a sense of pride on the Alberta side of the border stemming directly from the intentional and institutionalized nonexistence of rats there. That positive common identification in Alberta exists not because the prevailing values there are different from those on the Saskatchewan side, but because they are the same.

These commonalities notwithstanding, Alberta's rat control program gives the provincial boundary added significance that it would not otherwise carry by demarcating a specific form of difference.[2] In doing so, this spatial boundary becomes a significant landmark for Albertan cultural life by lending credibility to notions of a distinct provincial identity. It transforms what is otherwise an invisible line through the vast prairie into an important symbol in the cultural narrative establishing both collective identity and a sense of place. Though its cultural significance is primarily symbolic, especially given the cultural affinities shared by Albertans and Saskatchewanians in this area, the border's meaning is given salience by the physical distribution of the lives and bodies of rats. Likewise, rats themselves are symbolic cultural objects, but because they are also sentient animals, their cultural meaning is mobile. It is in the combination of our relationship to them in both these respects—symbolic and spatial— that rats become instrumental in the cultural narratives of boundary work. In other words, while rats' mere presence or absence lends a significance to the Alberta-Saskatchewan border, their symbolic

---

2   Notably, the distinction of being rat-free may generate a cultural difference that distinguishes Albertans along the border slightly from their Saskatchewanian counterparts. While Jesse and Phil cited a sense of shame surrounding rats posing difficulties on the Saskatchewan side, Jesse, Phil, and Dave all seemed to revel in the community cooperation that addressing one of Alberta's infrequent rat sightings requires. In fact, on multiple occasions they expressed mild disappointment that I had arrived just weeks after one of these had occurred and missed seeing this effort in action.

meaning as cultural objects is what makes this distinction important to collective identification and more than just an arbitrary difference.

THE RAT AS A (CONTESTED) CULTURAL OBJECT

While the cultural meaning ascribed to rats themselves is central to the broader meaning of the process of rat control, the semiotic construction of the rat in Alberta has a complex relationship to the real materiality of rat bodies. An awareness program in the early 1950s when the program was just beginning sought to educate Albertans on the visual appearance of rats, as many had never encountered one. The Department of Agriculture's annual report from 1952 states that "45 rat specimens, preserved in plastic containers, were placed in District Agriculturalists' offices and Schools of Agriculture to familiarize people with the appearance of rats" (Alberta Department of Agriculture 1952). This effort to make the rat an identifiable cultural object has encountered challenges at times, stemming from the necessity of drawing social, scientific, and cognitive boundaries around the category of "rat." In 1956, animals thought to be potential furbearers, referred to as "mura," were brought into the province, before being revealed to be "mutant Norway rats" (Alberta Department of Agriculture 1957). Initially, their owners were allowed to keep them, with special security provisions set for their enclosures, but by 1959 these owners had exterminated them (Alberta Department of Agriculture 1959).

Today, it remains important for the program that Albertans be familiar with rats and remain vigilant in case they see one. Besides the RCZ, the other main preoccupation of the program is systematizing response to rats that stow away on vehicles and materialize in other parts of the province. To do so, the program has a hotline (310-RATS) that Albertans can call if they suspect they have seen a rat, and the PCO of that area will investigate if the tip is deemed legitimate. However, issues of species identification persist to this day. For instance, during the three-hour drive from Calgary to the RCZ, a recent caller to the hotline sent Phil a picture via text message of a dead animal they suspected might be a rat. During a stop for gas, Merrill viewed the picture and made a swift judgment that this was in fact a picture of a muskrat and informed the caller. As he did so, he explained that this outcome—the potential rat sighting turns out to be another variety of rodent—occurs with the overwhelming majority of these calls to the hotline.

Callers have varying reactions when Phil communicates this determination. Some are happy to learn that Alberta's rat-free status is still intact. Occasionally, though, a caller will see the distinction between different varieties of rodents as trivial and still want a PCO to come assist with exterminating the rodent. As Phil explains, sometimes these other rodents might be causing property damage or other problems similar to what a rat would. Nonetheless, his response in these instances is to apologetically tell them "We have a *rat* program. We don't have a muskrat program." Ultimately, these instances show how the value placed on maintaining the province's "rat-free" status in and of itself for the program is on par with the practical reasons for rat control. While the damage caused by a muskrat or another similar rodent may be the same as the effects of rats, exterminating muskrats is not connected to a broader cultural narrative the way rat extermination is.

The program has also seen the limits of biological taxonomy as a criterion for the public's concept of the category of "rat" tested in a different way. Part of the public policy related to Alberta's rat control program is a province-wide ban on pet rats, to which a small population of would-be rat owners has objected. At question here is a set of criteria that extends beyond the formal definition of species for defining "the rat" in the Albertan cultural imagination. Pet rats are a different kind of subject than the "menace" and "pest" the government's program has constructed the animals as. For one, they do not draw the affective response (at least from their owners) that early propaganda posters produced by the program and Merrill's continuing awareness efforts suggest is appropriate for rats. They are voluntarily kept and cared for, rather than being considered "invasive" where they live. Furthermore, Merrill himself acknowledges that pet rats do not pose the same practical threats to his program's goals that the rats that might enter Alberta through the Saskatchewan border do, but he has reasons for keeping the ban in place: "People who have them as pets [say] 'why can't I keep a Norway rat?' and 'what if I neuter it?' and 'it's just a nice pet' and we got quite a bit of pressure from the pet owners who had pet rats. And we were concerned about that, not because their pet rat is one that's going to escape and [affect] our program, [but] because we want to try and teach everybody we're rat-free. And if everybody's got a pet rat, we're not."

Like the program's approach to non-rat reports to the 310-RATS hotline, this policy affirms the importance of biological taxonomy as a boundary around the category of "rat," in the interest of retaining the legitimacy of Alberta's claim to being rat-free. Nonetheless, Merrill's team deals with violations of this rule in a much different way than they deal with the other variety of rats they are concerned with. He tells me that his program has a partnership with an airline that flies the pet rats into neighboring British Columbia, which has no such ban on keeping them as pets, where they are delivered to a local shelter operated by the Society for the Prevention of Cruelty to Animals. So, while "species" remains the defining criterion guiding the province's overall anti-rat policy, there are factors, social in nature, that transcend this mode of classification and guide the program's specific practices with respect to the lives (and deaths) of the animals under their jurisdiction.

Another aspect of the program's approach to pet rats casts these contradictions into even starker relief. Merrill tells me that one of the ways his side has made peace with the population of pet rat advocates is with a compromise that allows pet shops to sell a species of pet "rat" called an African Soft Fur. Though this animal is commonly known as the "African Soft Furred Rat," taxonomically it is classified in the genus *Mastomys*, as opposed to *Rattus*, which is the classification that Alberta's program has in mind when they claim to be "rat-free." Thus, there are two notions of what constitutes a "rat" existing simultaneously, if somewhat uneasily, in Alberta: the formal scientific one favored by the rat control program and the more informal, commonsense one that residents interested in keeping a rat as a pet might adhere to.

In sum, the cultural narrative of rat control rests on a notion of the rat as a cultural object with a shared and understood meaning. The precarious relationship between this meaning and the lives and bodies of the animals themselves recalls similar observations regarding other cultural objects, such as McDonnell's examination of AIDS campaigns in Ghana, where the materiality of objects like awareness ribbons, billboards, and condoms continually thwarts the meanings and uses intended by the programs that produced them (2010, 2016), or Domínguez Rubio's work on the fickle and fragile nature of art objects (2014, 2016). While the explicit logic of rat control in Alberta is based in scientific taxonomy, for the purposes of cultural narrative it is ultimately a more malleable, socially constructed category.

As we will explore in greater detail in chapter 3, the construction of the rat specifically as killable and worthy of systematic eradication is essential to the process of placemaking and boundary work achieved through the material practice of rat control described in the preceding section. The claim of "rat-free" status and its significance rests on a collective, shared understanding of the boundaries around the category of "rat." Just as importantly, it necessitates that this shared meaning constructs the rat as a symbolic villain worthy of lethal force, which makes the spatial distinction marked by the geographical border a significant rather than arbitrary one and makes the idea of a "rat-free" province an achievement of moral superiority.

INEQUALITIES AND SYMBOLIC BOUNDARIES

Alberta's rat control program reveals how animals can play an important role in the drawing of boundaries around social categories and group identities. In this chapter, I discussed how the province's rat control program is a salient cultural narrative of placemaking and regional identification based on the combination of rats' symbolic meaning and humans' material, spatial interactions with them. Historically, Alberta's collective identification has been characterized by a potent cultural current of opposition to outside influence from the Canadian government and the cultural elite in the eastern part of the country and a related assertion of Alberta as geographically, politically, and culturally distinct. As discussed, this is complicated by the fact that, comparatively, Alberta has a greater affinity with nearby western provinces like Saskatchewan. Rat control provides a symbolic "outsider" against which the collective identification can be defined. The material process of rat inspection in the east of the province makes the provincial border a meaningful socio-spatial boundary of the province itself and of the province's cultural identity. The meaning of this spatial process, though, is animated by the symbolic cultural meaning of the rat itself. This relationship reveals the unique character of animals as both objects of material culture that we interact with in space and symbolic characters with meaning that transcends those immediate physical interactions.

Animals act as key vectors of symbolic meaning where spatial and symbolic logics consolidate into notions of "us" and "them." As we will explore in greater detail in the following chapter, the implications of these lines of demarcation are not value-neutral. Animals often become particularly instrumental in the more reactionary imaginations of place rooted in exclusionary politics described by scholars like Harvey (1997). In the United States, for example, one of the most divisive political issues of recent years has been the debate over immigration from Latin America and former president Trump's attempts to erect a national border wall. That proposed structure has taken on as much cultural meaning as it has geopolitical significance. Within the discourse around this issue, we have seen the language of animality leveraged a number of times, including Donald Trump's insistence that would-be immigrants are "not people," but "animals" (Davis 2018), and his son Donald Trump Jr.'s comparison of the proposed wall along the border to those surrounding enclosures at zoos (Durando 2019). The *New York Times* has also reported on the reemergence of moral panic regarding wolves in Germany, which, similarly to the case discussed in this chapter, mirrors right-wing anxieties over immigration there (Bennhold 2019).

The process of boundary work introduced in this chapter, wherein a meaningful notion of Albertan collective identity coalesces, is no less inflected with nativist sentiments. How does Alberta's investment in keeping rats out map onto other exclusionary impulses? The following chapter turns to this question and explores how the policing of the Alberta border against rats is importantly connected to nativist currents that emerge elsewhere in Albertan cultural life.

✳ THREE ✳

"YOU CAN'T IGNORE THE RAT"

Guarding Alberta's Moral Character

Driving through Eastern Alberta during my trip to the Rat Control Zone, Jesse, the pest control officer in charge of this area, commented that his father often jokes about Canada's prime minister Justin Trudeau and Canada's Liberal Party more broadly. When I asked him how he felt about Trudeau himself, Jesse replied, flatly, "I hate him." He went on to elaborate by saying: "He only represents the urban parts of Canada. He has no idea what it's like for rural people. No idea. He's bringing more immigrants in and giving more jobs to immigrants instead of Canadians." Jesse sees Trudeau's government as failing to prioritize Canadians like those in this area of Eastern Alberta, particularly through its approach to immigration policy and asylum seekers. These misgivings about the Canadian prime minister sounded instantly familiar to me as an American observer. My ride-along with Jesse and Phil in the Rat Control Zone took place in the middle of Donald Trump's first presidency, coinciding with his attempt to deliver on his most prominent campaign promise, erecting a wall along America's southern border with Mexico to deter migrants from crossing into the United States. A few months later, Trump and other right-wing figures would stoke fears of a "migrant caravan" en route to that border in an attempt to lean on a tried-and-true "wedge" issue ahead of the 2018 midterm elections. The United States then endured a lengthy government shutdown that winter stemming from a political impasse over funding for the proposed border wall.

The context that this timing provided surely contributed to

how Jesse's comments struck me. After a day spent crisscrossing the boundary between Alberta and Saskatchewan and observing the work that goes into preventing the rats that we found on the eastern side of that line from breaching it, the discussion of immigration and the way Jesse's sentiments mirrored what I was used to hearing from right-wing talking heads in the United States felt especially charged. At the very least, this view of immigration and the sense of dissatisfaction with Trudeau's government falls in line with the oppositional nature of collective identification in Alberta discussed in the last chapter. When Jesse laments how disconnected Trudeau is, in his opinion, from "rural" Canadians, this complaint echoes long-standing sentiments in western Canada and especially Alberta that those who live there are defined by a rural, agricultural way of life that sets them apart from the more urban eastern provinces.

And, while the overt symbolic importance of these connections likely varies depending on who you ask, both rats, which might cause destruction and economic hardship for Albertan farmers, and immigrants, who Jesse and others viewed as threatening to usurp employment opportunities, are seen as external threats within this worldview.

To put things another way, the oppositional character of collective identification explored in the last chapter implies the potential of interlopers from outside the province to challenge this identity. While this makes rat control and notions of a rat-free province a resonant narrative for regional identity, it also makes rats a convenient analogue for other objects of nativist fears or anger. This chapter examines the past and present relationships between rat control in Alberta and other imagined interlopers. While the goal of a rat-free province has never been overtly motivated by nativist opposition to human outsider groups, it has, as a narrative, always enjoyed resonance thanks in part to Albertans' concerns about other external threats to prevailing notions of the province's cultural identity.

The rhetorical connections between the narrative of a rat-free Alberta and racism or other forms of out-group animosity are occasionally more than vague, implicit suggestions in the realm of innuendo. In August 2022, the Canadian news outlet Global News reported that a Calgary-based neo-Nazi group had claimed responsibility for hanging a banner from a pedestrian overpass on McLeod Trail, a major Calgary thoroughfare. The banner contained a phrase known as the "14 words," which the Southern Poverty Law Center describes as a "rallying cry for militant white nationalists." The phrase reads: "We must secure the existence of our people and a future for white children." Among its online materials, interspersed with quotations from Adolf Hitler and other white supremacists, the group made this declaration: "We stand for a rat-free Alberta and operate on our own terms, as free Aryan men." This fascist group's invocation of Alberta's anti-rat crusade is a stark demonstration of the resonance it carries with exclusionary politics and its extreme possibilities.

The mutually reinforcing relationship between narratives of a rat-free province and narratives that villainize outsiders like potential immigrants highlights the role of different varieties of symbolic hierarchies. The construction of the rat as a symbolic villain generates a hierarchy of species ranging from humans on one end to the most killable of animal others like rats on the opposite end. Negotiating the boundaries of group identity, such as the collective identity of Alberta, means not only establishing a set of cultural values and iconography associated with a group, but also excluding others from accessing this membership on the basis of that identity. Who and what, besides rats, has Alberta historically sought to exclude from its cultural identity?

By exploring this question, I aim to glean broader insights into how the relationship between interspecies hierarchies like the one rat extermination rests on connects to the social hierarchies that form the basis of group identification. I argue that our dealings with nonhuman others are mutually reinforcing in this regard. In circumstances where boundary work involves exclusionary politics, it is more likely that narratives that villainize certain nonhumans will gain traction. Moreover, when they do, they form a reservoir of symbolic meaning that might reinforce the cultural prominence of nativism.

To begin, this chapter discusses the tensions that exist in Alberta today that affect the continuing prospects of the rat control program. The bulk of the rest of the chapter explores the rat program's early years in terms of both the program itself and the political and cultural environment in Alberta at the time. This historical look at the early days of rat control reveals how the goal of a rat-free province resonated with the cultural narratives of Albertan identity favored by the political regime that ran the province at the time. I then return to the present day to examine how the endurance of the rat control program interacts with modern social conflicts and struggles over collective identity.

THE CONTEMPORARY TENSIONS OF RAT CONTROL

One of the most notable features of Alberta's rat control program is its longevity. Since these government efforts began in the middle of the twentieth century, the province has been able to maintain its rat-free status (with some qualifications noted in the previous chapter) for several decades. Given that part of the strength of this program is its resonance as a narrative of collective identity, though, the continuity of the program has faced challenges in recent years stemming from shifts in the province's population that have made it more heterogeneous. When notions of collective provincial identity become less settled, in other words, the cultural meanings of rat control, rats themselves, and a rat-free Alberta are similarly disrupted. Alberta's demographics are shifting due to immigration, and the province is also becoming increasingly urban, which threatens the primacy of white rural culture as a boundary of regional identity. These changes accordingly challenge the resonance of rat control as a cultural narrative and its efficacy in the project of boundary work. Such dynamics were top of mind when I asked Phil about the biggest difficulties that his program faces: "I'd say the biggest challenge right now is . . . to keep Albertans educated [about] the program. . . . Because I think our—not our farming community so much—but our urban community is totally changing and that's where the votes are. . . . To keep them excited about the rat program is probably our biggest challenge."

The changes Phil notes are reflected in Canadian census data on Alberta's population. The population of Calgary, the province's largest metropolitan area, grew by more than 25 percent between 2006 and 2016, from 988,193 to 1,239,220 (Canadian Census 2006, 2016a). As this urban population grows, Alberta's racial and ethnic demographics are changing as well, with the last ten years seeing increased diversity in this respect. The percentage of the province's total population considered a "visible minority"[1] has increased from 13.9 in 2006 to 23.5 in 2016 (Canadian Census 2016b; Treasury Board and Finance Office of Statistics and Information-Demography 2017).

For Phil, these changes are consequential because they mean Alberta's cultural values are becoming less specifically tied to agricultural life. Particularly, he says the less connection Albertans have to rural life, the less enthusiasm they have about programs like rat control: "I think it's . . . a concern that these guys don't know where milk comes from, don't realize that somebody has to milk every morning, and therefore won't support those [types] of programs. They won't support the guys that inspect their milk and . . . [say] 'Oh, what are we spending money on that for?' And they don't understand."

Phil acknowledges some amount of tension between the more quintessentially "Albertan" culture of rural, agricultural life that has historically supported his program and the newer elements of Alberta that are more urban and more ethnically diverse. On a practical level, he worries that these demographic changes will mean the program's public support will decline to the point where it will lose its funding. Symbolically, this struggle reflects a crisis of the boundaries of identity in Alberta, as popular notions of regional culture become vulnerable to losing their salience. As a narrative of boundary work, the rat control program becomes a prism through which these tensions and anxieties are refracted, and the rat itself becomes a cultural object whose symbolic meaning is in flux. Various forms of institutional messaging have historically helped breathe life into the cultural narratives that make rats salient objects in Alberta, despite their absence from the landscape. Phil wants to reinvigorate that cultural messaging by "educating" Albertans who are not familiar with or enthusiastic about the program, so they become supporters, a goal

---

1   The term "visible minority" is a census classification derived from Canada's Employment Equity Act, and refers to "persons, other than Aboriginal peoples, who are non-Caucasian in race or non-white in colour."

which he acknowledges has been difficult: "[It's] not easy. . . . I'm having a really hard time even suggesting in the elementary school program, to give us an option, if a teacher wants to teach something on rats, 'here's a program.'"

For years, he tells me, children were taught about rats and Alberta's rat-free status in elementary school. This was part of a broader campaign funded by the provincial Agricultural Department to increase awareness of and enthusiasm for the program and, in doing so, give rats a specific meaning in the cultural lives of Albertans. These contemporary challenges and the nostalgia Merrill has for past years when rat control efforts enjoyed more robust public messaging point to the importance of the early period of the program for ensuring its long-term success. Moreover, understanding the role of the rat as a cultural symbol in Alberta requires comprehending the social and cultural context of the program's inception. What did the early years of the rat control program look like and what was the cultural and political context in Alberta within which it succeeded?

THE ORIGINS OF RAT CONTROL IN ALBERTA

Years before rats were first reported in Alberta, the province's government laid the groundwork for what later became its rat control program. In 1942, the province enacted the Agricultural Pests Act, which carried many stipulations that would be relevant to the fight against rats a decade later. The act established a set of powers and procedures relating to pests and pest control that, importantly, gave Alberta's Minister of Agriculture the authority to dictate what was considered a pest in the first place. Specifically, the legislation decrees that a "pest" is defined as "any animal, insect or disease which may from time to time be declared by ministerial order to be a pest and to be likely to be destructive of, or dangerous to any crop or to any live stock." Another important feature of this act was the responsibility it laid upon farmers and property owners for the task of pest control itself. It stated, for instance, that "every person shall take active measures to destroy all pests upon any land or other premises owned, occupied or controlled by him, and when given notice in writing by an officer directing the destruction of pests, he shall obey the notice." Beyond this, it also endowed government officers with the power to intervene in the case of inadequate compliance with pest control goals and established fines that could be imposed on property owners in these cases.

The Agricultural Pests Act is relevant for the rat control program that came later for two reasons. Its obvious practical importance, of course, is that it established the procedural and legal framework within which the program would operate. Second, though, its emphasis on the responsibility of individual farmers in these procedures treated the goal of pest control as a collective effort, a shared priority uniting the agricultural backbone of the province. In other words, years before a rat-free Alberta became a point of pride linked to the province's shared identity, the government treated pest control more broadly as a concern for all Albertans, not just specialized officers.

In 1950, rats were reported on an Alberta farm just west of the town of Alsask, which is located along the Saskatchewan side of the two provinces' border (Anon n.d.). Initially, officials' main concern related to rats stemmed from the public health threat they might have posed to Albertans as disease vectors. When the provincial government first took action to stop the westward expansion of rats and seal off the border against them, it was the Department of Health that oversaw this effort. Soon, though, that responsibility was transferred to the Department of Agriculture, which invoked the Agricultural Pests Act to formally declare rats a threat to the economic livelihood of Alberta's farming sector.

The successful effort that followed involved a mobilization of manpower and resources such as poisoned bait and traps, as well as administrative decisions like the establishment of the six-mile-wide "Rat Control Zone" along the Saskatchewan border that still exists today. Crucially, though, it also depended on an awareness campaign that informed Albertans about rats and rat control methods. In the process, this campaign elevated "the rat" from the physical bodies and lives of the rodents themselves into a salient cultural artifact that has meaning for Albertans. As noted earlier, many Albertans today have never encountered a rat and still often confuse other rodent species for them. This was even more of an issue in the early days. Beyond simply raising awareness, though, these informational efforts sought to build enthusiasm for and collective pride in the effort of making Alberta rat-free. To this effect, the province produced a series of propaganda posters in the 1950s to build support for the program. These materials promoted a fierce antagonism to rats, and seemingly attempted to engender a negative affective response to the sight or even the thought of the rodents. Alberta's Department of Agriculture described the poster campaign this way in 1951: "Over 2,000 large posters regarding rats were printed and distributed to elevator agents, railway station agents, schools, and post offices for posting in conspicuous places to inform the public of the encroaching menace" (Alberta Department of Agriculture 1951).

The posters themselves are far from subtle (see figs. 3.1 and 3.2). One poster reads "You Can't Ignore the RAT," and at the bottom adds "KILL HIM! LET'S KEEP ALBERTA RAT-FREE." Another depicts rats encroaching on the illuminated borders of Alberta, helping brand rats as the "menace" threatening the province from outside that the Department of Agriculture's report deemed them. The heading "Kill Rats With Warfarin," used on one poster, serves to strengthen the mental association of rats with extermination, by naming the primary anticoagulant poison used at the time (Alberta Department of Agriculture 1954).

Another indication of the tenor of this cultural messaging shows up at the bottom of the poster in figure 3.1. Text in the footer of the poster indicates it was commissioned by the provincial Department of Public Health's Division of Entomology. Entomology, of course, is the study of insects, and public health entomology refers to the study of insects' capacity to act as vectors for infectious diseases. This may be dismissed as simply an administrative quirk—when I asked Phil Merrill about this in an interview, he speculated that this corner of the Department of Public Health was simply "where the money was"—but it is nonetheless a meaningful and illustrative detail. It is almost always easier for the average person to accept and justify the killing of insects, compared with mammals. Moreover, in other contexts, such as medical research, rats even serve as direct proxies for human beings. But rather than call attention to the affinities we as humans might share with these animals, Alberta's rat control effort stood to benefit from discursively lowering rats' status from rodent

FIGURE 3.1 Promotional poster from the early years of Alberta's rat control program. Text at the bottom indicates the poster was produced by the Alberta Department of Public Health's "Division of Entomology." Image provided by the Provincial Archives of Alberta.

to insect, which further serves to render them, as cultural objects, dispensable beings worthy of lethal force.

In short, the Alberta government's public messaging during its campaign to rid the province of rats sought to drum up public support for this goal by crafting a specific cultural narrative about the animals. This narrative cast them as a malicious invading "menace" that threatened the collective well-being of Alberta. In the process, the notion that it was ordinary Albertans' duty to aid in the effort to identify and exterminate rats appealed to a sense of shared identity

FIGURE 3.2 Promotional poster from the early years of Alberta's rat control program depicting a horde of rats looming outside Alberta's borders on a minimalist map of North America. Like many of the program's early promotional materials, this image casts rats as a nefarious, external force threatening the well-being of the province. Image provided by the Provincial Archives of Alberta.

in the province. What made this an effective rhetorical strategy in 1950s Alberta?

ERNEST MANNING'S ALBERTA

The first decade of Alberta's rat control program came in the middle of the twenty-five-year service of Ernest Manning, the longest-serving premier in the history of the province. Manning's term of office has had a lasting legacy in Albertan politics and the right-wing populism that has flourished there since. In fact, Manning's son Preston founded the populist Reform Party that shook up the national political landscape in Canada during the late 1980s and 1990s by attempting to carve out a greater share of power for the country's western provinces. The elder Manning headed up Alberta's government for a period that stretched from the middle of World War II to 1968, during which his Social Credit Party came to dominate provincial politics.

The values and rhetoric of Manning's administration shed light on the broader cultural context that provided the backdrop for the rat control program's inception. The overlap between the rat control program and the Manning premiership was a time in which Alberta was in the process of consolidating a shared cultural and political identity. The year 1955 in particular presented an opportunity for reflection on this identity, the province's history, and the set of values that defined Alberta as a collective whole. That year Alberta celebrated its Golden Jubilee, the fiftieth anniversary of its attainment of provincial status in 1905. Ernest Manning's public addresses during this year offer a useful window into his imagination of Albertan history and values. In one address marking this anniversary, Ernest Manning tells a sort of origin story for the province defined by the industrious contributions of intrepid pioneers: "In the Alberta of 1905, the vast natural resources remained untapped while the people concentrated their energy and resourcefulness to the task of building a productive agricultural empire. In this Golden Jubilee year, our province has achieved an agricultural ambition which, together with the development of natural resources, is broadening the industrial base of the Province and widening economic opportunity" (Manning 1953–1955a).[2]

Manning's telling of Alberta's past and present emphasizes the agricultural sector and a more general taming of the natural world to harness its material resources as both integral to the province's identity and key to its future success. Moreover, he sees Alberta as continuing in the footsteps of Albertans past, whom he regards as having transformed the province from untamed wilderness to industrial

---

2   Ernest Manning, Premier Papers—Addresses and Speeches, 1953–1955, PR1969.0289/1826. 40.72, Provincial Archives of Alberta.

powerhouse. Specifically, he uses the Golden Jubilee as a chance to "honor our pioneer citizens who laid the foundations of our present attainments by their courage and toils." Elsewhere, Manning strikes an even more laudatory tone in celebrating the first white settlers of Alberta. In a radio address, he poses the rhetorical question of how Alberta became a "thriving" province, before declaring: "More than any other single factor, it was the pioneers—those farsighted men and women who persisted in the undertakings until they saw a great land shaping beneath their toiling hands. It was the tireless work of fur traders, of missionaries, homesteaders, ranchers, and the countless others who blased the traile for those who followed after them."

Tributes like these to the white homesteaders that Manning sees as having tapped the "virgin" land's potential are the most prominent theme in his addresses marking the Jubilee. While this glowing narrative of industriousness and hard work in the face of adversity is itself an indication of the values that Manning saw as defining Alberta's collective character, he also commented more specifically in these addresses on how he saw contemporary Alberta taking up this mantle. One aspect of this was the enduring trial posed by the province's harsh landscape and the will of Albertans to continue to overcome the obstacles it posed while taming that landscape. Commenting on a more recent generation of Albertans, for instance, he struck an elegiac tone about hardship and endurance: "They saw the bitter lean years of the Depression; they saw prairie winds carry away the rich topsoil in black clouds of dust; they chopped holes in the ice of streams and lakes to water their freezing cattle; they searched for newborn calves in late spring storms; they huddled close to stoves while blizzards raged across their homesteads. These people are not pioneers of the dim past; they are people who are still among us."

Given this emphasis on the triumph of hard work and resilience over the challenges of nature, one way that the fight against rats might fit into overarching narratives of Albertan character becomes clear. The march of rats across the Canadian Shield might be seen as yet another trial posed by the natural world that Alberta would have to summon hardiness and determination to overcome, the same way it does with blizzards and Chinook winds. This factor likely did contribute to making rats a particularly resonant symbol in the province during this period. The thorough symbolic construction of the rat as a villain bent on spoiling Alberta's success and well-being, though, made the rodents something more than just a challenge posed by nature. Other relevant elements of Albertan cultural identity, as imagined by Manning's administration, provide useful context for this moralized rhetoric about rats.

The Moral Character of Alberta and Its Vulnerabilities

For Manning, the industriousness of the pioneers about whom he spoke so glowingly in his Jubilee addresses was firmly rooted in Christian faith, which he in turn held to be a core aspect of Alberta's collective identity. He prefaces one of his lengthy tributes to these homesteaders by declaring: "They have produced a province of honest, God-fearing citizens who are proud of their accomplishments which now constitute an impressive heritage" (Manning 1953–1955a). In another address delivered to Alberta schoolchildren, Manning emphasized the "Christian character" required to build a "worthwhile community" (Manning 1953–1955a).

For Manning and his ilk, the importance of Christian faith extended beyond the idolization of missionaries who were among the early white settlers of Alberta. The notion that Christian morality is a central bedrock of the province is illustrated by comments he made in a 1953 issue of the *Calgary Herald*, which resulted in mild controversy. In these comments he took a swipe at the moral character of the United Kingdom by challenging that country's religiosity, remarking that "Less than 10 per cent of the British population attend churches... but the dog races are followed by thousands" (Manning 1953–1955a). While some Albertans took issue with this statement (the Provincial Archives of Alberta has in its collected documents related to Manning multiple letters to the premier noting that "there is nothing criminal in attending dog races," among other objections), the original quotation from Manning indicates the sort of moral worldview that Manning and his administration wished for Alberta to uphold. His disparagement of the UK's moral character on these grounds, meanwhile, serves to highlight the supposed righteousness of Albertan culture by contrast. Manning was also not shy about the fact that he saw his political career as a sort of religious calling. In a 1950 address to the National Gideon Convention, he repudiated criticisms of this notion:

> As I travel around this country I frequently meet those who think it strange that a man in public life should be interested in the Bible and in trying to talk to men and women about the importance of their relationship to Jesus Christ. I sometimes wonder why people should think it strange. In public life we are called upon to grapple with the complex problems of human relationships which stem from man's inhumanity to man. If you are a realist, you cannot deny the fact that beneath and behind all of our human and material problems there lies one basic problem of man's broken relationship to his God.

Manning's idea of Alberta's cultural identity is grounded in the bootstrapping narratives of hard work and determination exemplified by hardy homesteaders, but even more than that he views Albertans as a righteous Christian people. In that same address he describes how hearing a radio proselytizer from Calgary prompted his journey from a humble farm in Saskatchewan to the calling of spreading Christian values in his role as a statesman.

These details are all to note that there was a very specific moral worldview informing Manning and his Social Credit Party's leadership of Alberta. Equally important to this moral worldview, though, were the threats that Albertans like Manning envisioned to it at this time. While he may have had disdain for the vices indulged at British dog tracks, his greater worry was farther to the east. Manning was a staunch anticommunist who fiercely opposed labor organizing in Alberta on these grounds. He also gestures to what he perceives as a moral depravity of communism and the imminent threat it poses in the same address to the Gideon Convention: "Consider for a moment the significant trend in world events today. Over one-third of the people of this earth already are under the domination of men whose philosophy of life repudiates God and throws into the discard His infallible Word. The accumulative consequences of human depravity have pyramided down the centuries with each successive generation until they now are precipitating a crisis unparalleled in history" (Manning 1953–1955a).

While concern over the geopolitical balance of power and the growing influence of the communist bloc were of course widespread in the Western world at this time, Manning was deeply concerned about the possible erosion of Christian values at home. In comments provided for the Alberta Council on Family and Child Welfare's annual publication, he framed this possibility as not an external threat, but a more insidious internal problem that implored Alberta to keep moral vigilance: "A nation is only as strong as the stability of its home and family life. As a result, the greatest single threat to our Christian civilization today does not come, as many believe, from behind the Iron Curtain but, rather, it lies in the progressive disintegration of home and family life within our home country" (Manning 1953–1955b).

Despite this particular phrasing, however, Manning's government was concerned about threats to Alberta that came from "behind the Iron Curtain" as well. In an address on the importance of civil defense in both Alberta and Canada at large, Manning took seriously the possibility that his province might be the target of a nuclear attack. While he acknowledges that the more likely targets of such a strike were elsewhere in North America, especially the urban centers of the United States, he nonetheless envisioned many possible scenarios where the USSR might threaten Alberta specifically. For instance, he noted: "Our large cities such as Edmonton, Calgary, and Lethbridge may be rated as possible targets by the Russians—we don't know. We do know that Alberta lies in the broad flight path between North East Asia and the large industrial areas in the central part of this continent. Therefore, in any attack on the central parts of North America, it can be expected that enemy aircraft would attempt to pass over Alberta. It is known that the Russians have an aircraft similar to the American B-29 which is capable of delivering atomic and other heavy bombs."

Manning also saw Alberta's geographic position as posing further potential dangers. After conceding that Russian fighter jets that entered North American airspace would likely head for "larger centres in the US," he discusses how this fact should be of no comfort to the average Albertan: "There will be times, however, when our fighter aircraft will turn them back . . . anti-aircraft gunfire and bad weather will turn others back. In such cases the enemy planes may be turned back over Alberta when they are still carrying their bombs. Before returning home or bailing out the enemy crews will be anxious to drop their bomb-loads on some convenient target. These conditions will bring some of the dangers from air attack not only to our larger cities but to our smaller towns and villages too" (Manning 1953–1955a).

The genuine concern that Manning's administration had regarding a possible attack on Alberta highlights its broader preoccupation with communism as a threat to the province's well-being. Of course, the more acute danger imagined by the socially conservative Alberta government was likely the internal rise of communist ideology, especially through organized labor, rather than nuclear bombs raining from the sky. Nonetheless, the fear that Alberta might be the target of Russian attack reinforces the notion of a unified, morally righteous Albertan culture that must defend itself against external threats.

This adds context to the snarling rats depicted on propaganda posters in this same era as looming menacingly along the borders of Alberta. Alberta's anti-rat campaign was not overtly designed to be a symbolic, anticommunist proxy war. However, the fear of communism and the contemporaneous campaign against rats fit together harmoniously as pieces of the broader Albertan identity, as imagined by its government. In a semiotic analysis of Alberta's rat control posters from this era, McTavish and Zheng (2011) argue that this propaganda campaign promoted the idea of Alberta as a unified people characterized by their strong work ethic and cleanliness. The material fight to keep Alberta rat-free, to keep productive Albertan farms unspoiled by the rodents, resonates with the simultaneous goal of resisting the moral despoilation that communism might bring to Albertan culture. The values that the socially conservative Manning administration ascribed to Alberta's collective identity—industriousness, Christian morality, hardiness—are solidified through defense against external threats. In other words, the goal of vanquishing the rat as a personified villain and the notion that a nuclear Russia might attempt to impose a diametrically opposite set of values with an aerial attack both resonated with the broader oppositional nature of Albertan identity.

Alberta's Crusade against "Feeble Mindedness"

The public rhetoric from Ernest Manning's addresses examined above paints communist ideology as a key threat to Alberta's moral character. While the Red Scare looms as a cultural touchstone of the 1950s, the Alberta government had another, less public front during this time in its efforts to guard notions of moral "purity" for its populace. Between 1928 and 1972, Alberta maintained a forced sterilization program that was one of the longest-lasting such programs in North America. Legislation referred to as the "Sexual Sterilization Act" (1928) empowered a Eugenics Board to recommend the sterilization of Alberta citizens on the basis of "feeble mindedness" (McLaren 1990). In the early period of this program, the sterilization procedure required the consent of the patient or a guardian. However, under the government of William Aberhart, the founder of Manning's Social Credit Party and his predecessor as premier, Alberta removed this stipulation from the act and allowed for compulsory sterilization as recommended by the Eugenics Board ("An Amendment to the Sexual Sterilization Act" 1942). The 1950s and 1960s would become the most active period of this program (McLaren 1990).

This dark period in Alberta's history had a public reckoning in the 1990s when a woman, Leilani Muir, successfully sued the province for $2.5 million for her forced sterilization, which occurred forty years prior and without her knowledge or consent (Wahlsten 1997). Muir was one of over 2,800 people sterilized between the passage of the Sexual Sterilization Act and its repeal (Cairney 1996; Grekul, Krahn, and Odynak 2004). Like many others among this total, Muir was admitted to a state-run facility for "Mental Defectives." Among the collected documents related to Ernest Manning's premiership at Alberta's Provincial Archives is a list of health care–related initiatives to be implemented imminently, which contains an ominous record of the sterilization program. Following action items such as a program for the provision of expensive, lifesaving medications, a preventive program for blindness, and the planning of a Calgary hospital intended to "provide diagnostic service for Albertans second to none on the continent," the last two items call for facilities in Edmonton and Deerhome to house "mentally defective" and "severe congenitally deformed" children. It was this type of facility that presented cases to the Eugenics Board for possible sterilization.

While the practice of sterilization in Alberta was motivated by the goal of reducing "mental defectiveness," this rationale allowed for broad interpretations and applications that amounted to the enforcement of moral values. In explaining the longevity of Alberta's sterilization program, especially given the fact that many similar programs were terminated after the world learned of the horrors of Nazi Germany's eugenics program, sociologist Jana Grekul points to the appetite of the province's Social Credit regime for this kind of moral enforcement, especially in light of changing gender roles (2008). She describes the program as taking a "two-pronged" approach, which carried out more traditional eugenicist population control aimed at cleansing the gene pool of perceived congenital deficiencies alongside sterilization on the basis of "moral and deviant behaviors, intricately connected to sexuality, and particularly as they relate to females" (Grekul 2008). The sexual activity of women presented to the board for possible sterilization was heavily scrutinized and provided the rationale for many of these decisions. Based on her analysis of these case files and considering the context of Alberta under Ernest Manning, Grekul concludes: "In the politically conservative province of Alberta, run by an authoritarian Social Credit regime for four decades, led by premiers who were also fundamentalist religious leaders it seems plausible that the province's mental health professionals would and could draw on sexual sterilization legislation in an effort to control the mentally normal women who were challenging gender norms and the establishment."

Beyond the enforcement of a set of conservative moral values around sexuality, the history of eugenics in Alberta and Canada at large is inextricably linked to racist anxieties about immigrant populations. In the 1890s and 1900s, Canada saw a surge in immigration from non-Anglo-Saxon origins. The early push for the implementation of eugenics in Canada was amplified by figures like J. S. Woodsworth, a reformer of the Social Gospel movement who bemoaned what he perceived as the inferior character of new arrivals in the country. For instance, as Angus McLaren notes in his history of eugenics in Canada, Woodsworth drew distinctions between what he viewed as favorable and unfavorable populations of immigrants: "He contrasted the Scandinavians and Icelanders ('clean-bodied' and 'serious-minded as a race') to the Slavs and Galicians ('addicted to drunken sprees' and 'animalized')" (McLaren 1990). Moreover, the concern about "feeble mindedness" that fueled the eugenics programs in Canada that followed was highly racialized: "At the 1914 Social Service Congress of Canada conference Helen MacMurchy rose to declare that the problem of defective children could only be solved if special education and medical inspection were complimented by restriction of education. 'It is well known to every intelligent Canadian,' she asserted, 'that the number of recent immigrants who drift into institutions for the neuropathic, the feeble-minded and the insane is very great.'" While this quotation from MacMurchy, a prominent Ontario-based eugenicist, shows how these nativist views were present across Canada, the institutionalization of eugenics in Alberta was hastened by the particular fervor of anti-immigrant views there and the higher proportion of immigrants that settled in the prairie provinces (McLaren 1990).

Whether it was combating perceived cultural and physiological inferiorities of non-Anglo-Saxon immigrants or policing the sexuality of women, the forced sterilization program of Alberta acted as a mechanism for protecting notions of moral purity in Alberta. Such notions were a prominent facet of the version of Alberta's collective identity endorsed by the Social Credit regime in the mid-twentieth century. Ernest Manning's public rhetoric paints a cultural mythology of Alberta as a productive agricultural empire continuing a tradition of Christian values forged by early white homesteaders and missionaries. This version of Alberta's identity omits the heritage and contributions of indigenous Albertans, non-Anglo-Saxon immigrants, and non-Christians. Moreover, it is clear that Manning's Social Credit regime was concerned about both internal and external threats to the moral purity of Albertan identity. One such threat was communist ideology that might arrive in Alberta by military force or more insidiously through organized labor, and in the process, hamper the province's capitalist economic model or dilute its religious morality. Another threat was Albertans themselves who either did not fit the vision of Albertan identity conjured in Manning's speeches or actively challenged the moral values that defined it. The sterilization program, which disproportionately targeted indigenous peoples and non-Anglo ethnic groups and enforced conservative sexual politics (Grekul 2008; Grekul et al. 2004), underscored the exclusivity and inequality built into this moral order.

The enforcement of a moralized notion of Albertan identity, whether though anticommunism or eugenics, adds crucial context for understanding the beginnings of Alberta's rat control program. For one, the notion of a "rat-free" province resonates with broader concerns in this period over moral purity. The vehemence on display in the promotional materials for the rat control program, which crafted the rat as a symbolic villain bent on invading Alberta, makes sense as a rhetorical strategy within this cultural environment. The worry over rats' potential to stymie agricultural production should they arrive in Alberta is reminiscent of the nativist concerns of eugenicists over the "declining quality" (Grekul et al. 2004) of immigrant populations who may not share the same work ethic, grit, and determination ascribed to the pioneers that Manning celebrated. The project of making and maintaining a rat-free Alberta was a cultural narrative that fit into the larger identity-crafting on display in Manning's speeches.

The moralized notion of Albertan identity that the province's leadership imagined in the 1950s is firmly rooted in social inequalities. The sterilization program that operated during this period highlights how that inequality was operationalized in a real, material way, where the medicalized notion of "defectiveness" and the power to prevent reproduction solidified social hierarchies. Similarly, the use of systematic lethal force against rats and the propaganda posters personifying them as malicious villains evoked a symbolic hierarchy of species that places rats near the bottom. Put another way, the promotional materials of the rat program's early years show how our social categorization of animals, which is more often thought of in scientific rational terms of biological taxonomy, is instead more fluid and operates similarly to the process of racial formation as described by Omi and Winant

(2014), wherein categories are given social meaning according to the construction and maintenance of a hierarchical order. For example, the discursive devaluation of rats implied by their inclusion in the jurisdiction of "entomology" is strikingly similar to the way Pellow (2017) explains how an assumed hierarchy of species allows white supremacists to leverage language of "animality" to subjugate people of color (or, for that matter, how early eugenicists in Canada described certain immigrant communities as "animalized"). The construction of both these hierarchies in Alberta—social inequalities and the relative symbolic value of species—operated within the same system of cultural values and were thus mutually reinforcing.

RATS AND THE NATIVISM OF TODAY

The Social Credit Party of Ernest Manning's era no longer exists. Nor does the forced sterilization program described in the previous section. Worries of communism do not preoccupy Alberta's political leadership today like they did during the 1950s. But Alberta is still rat-free and pest control officers still patrol the Rat Control Zone to prevent them from crossing the Saskatchewan border. While the shifting demographics of the province and its increasing urbanization have unsettled the notion of a collective Albertan identity, many residents of the province, especially those with close ties to the agriculture industry, still view rat control as important and being rat-free as a point of pride. One caller to the 310-RATS hotline who I interviewed demonstrated the program's enduring connection to Albertan collective identity when I asked him whether his decision to call the number was motivated more by a desire to keep his personal property free of rats or to help with the broader effort of keeping the province rat-free. He replied, "Those two are the same thing, as far as I am concerned," indicating that controlling rats on his own property is connected to the province's larger project; as an Albertan, the goals of Alberta are his goals as well.

The longevity of the rat control program also suggests the endurance of notions of moral "purity" and freedom from outside influence as key boundaries around Albertan identification. The boundary work of this identification can be observed, for instance, in Alberta Agenda's rhetoric of establishing "firewalls" to protect the province from oppressive government influence discussed in the previous chapter. It is also present in the comments from Jesse on Canadian immigration policy, which suggest that some Albertans continue to hold negative views of new arrivals to Canada, echoing the early twentieth-century anxieties that fueled the rise of eugenics.

Evidence from surveys indicates that Jesse's comments reflect views that are more prevalent in Alberta than other parts of Canada. The Canadian survey research organization Environics conducted a nationwide survey on topics related to immigration in 2018. Respondents were asked to what degree they agreed with the following statements about immigration: "Overall, there is too much immigration to Canada"; "Many people claiming to be refugees are not actual refugees"; and "On balance, immigrants to Canada are making Canada a worse place." Binary logistic regression models, summarized in the table in appendix A, measure the effect of Alberta residence on the likelihood that a respondent agreed (either "somewhat" or "strongly") with these statements. The models found significant positive associations with Alberta residence and agreement with each of the three prompts. Albertan respondents were approximately 39.3 percent more likely compared to other Canadians to hold that there is too much immigration to Canada, 44.8 percent more likely to believe that many people claiming refugee status are not actually refugees, and more than twice as likely to believe that immigration makes Canada a worse place overall.[3]

Most likely, few Albertans consciously view rats as direct proxies for possible immigrants or other imagined human interlopers, or that the rat control program is a direct and causal driver of anti-immigrant

---

3   Models include control variables for sex, age, educational attainment, and the population density for the respondents' city. This last variable was included to account for possible urban-rural divides in sentiments about immigration. Values for population density were not included in the original dataset published by Environics but were instead computed by sampling GIS 2020 data published by Columbia University's Center for International Earth Science Information Network (SEDAC).

sentiments. Nonetheless, the endurance of nativist attitudes toward potential immigrants in Alberta is an indication of the broader cultural environment within which the notion of a rat-free province became a collective point of pride. Rat control functions not only as a material practice to secure Alberta from rats, but also as a cultural narrative of border maintenance, which generates meaning that might be accessed in other sites and mechanisms of boundary work, such as these salient currents of nativism. As discussed above, the cultural narrative of rat control is inscribed spatially on the landscape itself and reproduces itself by ascribing a human-constructed morality to the lives of rats.

Rat control in Alberta, while rooted in a particular formulation of interspecies hierarchy, contributes to narratively laying the groundwork for power relations elsewhere in social life based on similarly conceived insider-outsider relationships. Likewise, nativist politics in turn make the border policing narrative of rat control more resonant itself. The notion of a "rat-free" Alberta, in other words, is an actively maintained narrative that contributes to defining what it means to be "Albertan" more broadly, which in turn dictates who is or is not; at the same time that rats are marked as killable in Alberta and, as noted above, even discursively lowered to the status of insects, antagonism toward outsiders like potential immigrants generates a similarly structured moral hierarchy along the lines of group membership.

While the goals of rat control are not motivated by any overt nativist politics, they provide a convenient, ready-made metaphor for the likes of militant white supremacists, as mentioned in this chapter's introduction. Establishing a symbolic hierarchy that pointedly renders some species of nonhuman animals as killable is not a value-neutral process, in other words. On one hand, the specific character of the early years of Alberta's rat control is explained by the cultural environment of the time. A political regime that villainized internal and external threats to the particular moral identity it sought for the province was perhaps uniquely positioned to apply the same kind of moral narrative to an interspecies conflict with rats. The cultural implications of this, however, are not unidirectional. Decades later the regime that started the rat control program is long gone, but nativist politics that might lean on the program's symbolic overtones remain.

THE LESSONS OF ALBERTA'S EXCEPTIONALISM

Alberta's more than half a century of rat control offers a unique window into the role nonhuman animals can play in negotiating the boundaries of group identity. Rat control in Alberta fits neatly into long-standing cultural mythologies of the province and its collective character for a number of reasons that are both material and symbolic. On the material side, the fact that rats' survival hinges on the food sources and shelter that humans provide and that their success typically comes at a detriment to farming goals makes them resonant as a boogeyman for the agricultural province. Alberta's politics and the discourses around ideas of a collective provincial identity, meanwhile, have good use for such a symbolic enemy. The tendency for Alberta to define itself in opposition to the rest of Canada makes a struggle against an external invader like rats a generative cultural narrative. Meanwhile, the various forms of nativism that the province, its politics, and its notions of collective identity have incubated over the years make the idea of a "rat-free" province resonant with broader discourses of "purity."

The unique symbolic and spatial roles that rats play in boundary work, as explored in the preceding two chapters, are grounded in the logic of a moral hierarchy of species. A program like Alberta's requires a collective acceptance of the notion that rats can and should be killed. This places them at the bottom of a hierarchy that extends from humans at one end, to companion animals and charismatic wildlife below them, on down to the most killable of creatures. Alberta's informational campaign in the early going of its rat control program consciously sought to place rats at the bottom of this spectrum by imploring Albertans to kill them and lumping them in with insects by including them in the domain of entomology. Given the discursive connections between rat control and various forms of nativism explored in this chapter, this shows that the structure of this interspecies hierarchy is linked to specifically human social hierarchies. To be clear, this is not to imply anything close to a moral equivalence between the work of pest control officers and white supremacy or eugenics. Rather, these connections simply show that our dealings with nonhuman animals are never value-neutral, that they echo through the meaning systems that structure social life and back again.

The case of Alberta is illustrative of these points for its uniqueness. Its claim to being the largest intentionally rat-free area in the world makes for a somewhat extreme example of how cultural meaning might intersect with rat control. But while Alberta is exceptional for its rat-free status, the practice of rat control itself is mundane and widespread. In this respect, the subject of the next two chapters provides a contrast. Instead of an anomaly in the field of rat control, they explore a much more quintessential, everyday aspect of the phenomenon: rats in cities. The urban arena of rat control highlights rats' liminal status on the border between nature and society, an important aspect of their cultural meaning that is a throughline connecting the three case studies of this book. This key symbolic boundary is the nexus linking human and nonhuman forms of difference and inequality.

PART II

THE BORDERS OF URBAN NATURE

✕ FOUR ✕
RATS AND THE INDOORS/OUTDOORS DIVIDE

Most often, when people refer to Los Angeles's City Hall, they refer specifically to the thirty-two-story Art Deco tower that houses the mayor's office and the city council chambers, an iconic 1920s-era building that looms over the US 101 freeway in the heart of downtown. More accurately, though, City Hall is a complex of buildings that also includes a cluster of other nearby high-rises. The buildings themselves can at times feel as impenetrable and labyrinthian as the bureaucratic machine they contain. I learned this the first time I visited City Hall for this project, to interview an employee of the city's Department of General Services. I arrived comfortably early but then spent an hour locating the appropriate parking structure for a visitor (where I made the grave error of choosing an entrance intended for employees only) and then misinterpreting the directions of multiple city employees before arriving out of breath and barely in time for our scheduled meeting.

In February 2019, the *Los Angeles Times* and other news outlets reported that City Hall had a rat problem. Local news stations picked up amateur footage of panicked bureaucrats shrieking and recoiling as a rat scurries across a hallway, before one employee attempts to trap it in a cardboard box. These reports coincided with a documented typhus outbreak in downtown LA, and, when a city attorney caught the disease and subsequently sued the city, claiming she contracted it while at work, these two events became inextricable. In the aftermath of this incident, the LA City Council weighed various possible actions, including replacing all the carpets in the entirety of the City Hall complex. Around the same time, Herb Wesson Jr., the president of the LA City Council, released a statement that read, in part: "Employees shouldn't have to come to work worried about rodents. We will do whatever it is we need to solve the problem and protect our city's public servants and those who visit city hall."

When Wesson refers to workers being "worried about rodents," it can be taken in one sense at face value; he surely believes that city employees should be free from unpleasant close encounters with rats like those documented in the abovementioned video clip while going about their jobs. In a larger sense, though, Wesson was likely referring to the broader implications of these encounters. For employees in City Hall, to "worry about rodents" is also to worry about the fleas that those rodents may harbor and the diseases like typhus that might be transmitted by either of those unwanted nonhuman office mates. The infestation in City Hall served as an uncomfortable reminder that urban life, even under the fluorescent lights of a highrise office building, unfolds within a broader ecosystem. Nonetheless, it is worth noting again that Wesson refers specifically to rodents in his statement, which is in turn an indication of the power of rats not just to spread disease or inspire revulsion, but to symbolize all those various ills at once.

In examining the particular case of LA City Hall's rat problem and the city's response to it, this chapter explores how rats contribute to ordering our social and spatial experiences of city life through their potent symbolic meaning. In urban contexts, rats are living avatars of dirtiness and disease, as well as other social ills like poverty, and, as will be discussed further in the next chapter, homelessness. In this context, the sentiment that city bureaucrats should not have to "worry about rodents" has even more expansive symbolic undertones; concern over rats in City Hall is an indictment of LA's capacity to function in a fundamental way as a city. I argue that a guiding principle underpinning much basic city administration is the maintenance of what I refer to as the "indoors/outdoors divide." On one hand, rats may remind us of the multispecies ecologies that overlay urban areas, which concrete, glass, and asphalt often invite us to overlook. Rat *control*, though, means managing that ecosystem to keep nature in its appropriate place. Materially, this means protecting office workers and other Angelenos from the hazards associated with nature (in this case, exposure to vector-borne diseases like typhus). Symbolically, this means guarding public faith in the threshold between "indoors" and "outdoors," however illusory that boundary might be at times in its material reality.

When the rat problem in City Hall began, the city initiated a wide-reaching response that addressed these tandem goals. It included, among other measures, more frequent inspections performed by the city's contracted pest control vendors, a new protocol of weekly cleanings of the exterior of the buildings, and the removal of much of the landscaping in the area in order to make the environment less hospitable to rats. This last item highlights the particular importance of the indoors/outdoors divide relative to how we collectively imagine safe, functioning, and desirable urban environments. Landscaping, street trees, and parks are all various forms of urban nature that have come to be regarded as crucial to an idealized city life. While urban planners and city governments welcome nature into the city center on these terms, nature can come in other, more menacing forms like rats. Urban greening efforts improve quality of life and symbolize a healthy harmony with the nonhuman world. Rats, on the other hand, spread diseases and symbolize cracks in the efficacy of urban administration. LA's decision to remove the landscaping surrounding the Civic Center illustrates the tension that these "good" and "bad" forms of nature pose for city governments. The indoors/outdoors divide, I argue, becomes an organizing principle around which city officials manage that tension; by and large, this means embracing and cultivating "good" nature in the outdoors while preventing "bad" nature from breaching the indoors.

This chapter and the following one explore the notion of an indoors/outdoors divide and examine how certain cultural objects, like rats, perform unique symbolic work that makes this boundary meaningful and important for organizing city life. To that effect, I argue that rats are among a small category of living embodiments of nature who carry such potent symbolic meaning that they order our social experiences of urban space. We understand the spatial logics of cities in terms of ontological dualisms like nature-society, human-nonhuman, and especially the very basic notion of the threshold between "indoors" and "outdoors." By trespassing them, rats clarify these meaningful distinctions. In light of this, the city's rat control effort is not simply a project of managing the physical distribution of the animals, but is more broadly concerned with maintaining the particular terms of our social experiences of urban space.

GOOD AND BAD NATURE IN LA

The saga of City Hall's rat infestation is a case of Los Angeles grappling with the complex dynamics of nature in urban life. It is worth zooming out briefly to consider LA's own broader relationship with nature. To many, the very idea of nature in LA is oxymoronic. If there is a "nature/society divide," then surely the epitome of the society half of such a dualism would be a metropolis like Los Angeles that has exploded into a sprawling, suburban megacity that stretches, by some definitions, from San Clemente in Orange County to Thousand Oaks, some 100 miles to the north. Jenny Price, in her 2006 essay "Thirteen Ways of Seeing Nature in LA," wrote that the "reigning nature story" about Los Angeles is that nature is simply nonexistent there. This reputation—as a bastion of anti-ecological human enterprise—only heightens the ever-evolving tension between the human and nonhuman worlds in LA. Even if we accept the notion that there is no nature in LA (which we should not), we are forced to acknowledge the herculean efforts that went and continue to go into bending the natural world to humanity's will to make the statement even remotely plausible. Whether it be a massive Army Corps of Engineers project to prevent flash flooding by confining the volatile Los Angeles River to a "concrete straitjacket" (Price 2006), or decades of dubious fire suppression policy in the San Gabriel mountains that only worsened eventual wildfires (Davis 1998), LA is always contending with the natural world's habit of reminding Angelenos of the ecosystem that their city is built within.

Moreover, as countless scholars have already shown, the notion of a nature/society divide is at best an illusion. Cities are ever-unfolding socio-environmental processes (Heynen, Kaika, and Swyngedouw 2006). They not only depend on raw materials and resources from the natural world beyond, but they are also multispecies ecologies in their own right, supporting flora and fauna well beyond their human occupants. In a more symbolic way, urban nature is also central to urbanization. While common narratives cast urban nature like parks, street trees, and landscaping as a necessary remedy to urban ills like overcrowding and pollution, Hillary Angelo has argued convincingly that such forms of nature have in fact been a central part of the social imaginaries guiding urbanization itself, and key elements for crafting a desirable, recognizably urban area (2021). Of course, cultivating urban nature in a way that accomplishes that goal is a complex task in both material and symbolic terms. The nonhuman world is often less than cooperative with the human meaning systems we attempt to enroll it in, and urban greening may threaten to compromise other priorities of urban administration, such as public health.

P-22 as an LA Nature Story

LA's task of balancing these dynamics of nature is on display in another animal story chronicled in the *Los Angeles Times* that contains an echo of the City Hall rat problem. The protagonist of this story is P-22, a mountain lion that gained "celebrity" status after being spotted on a camera trap in 2012 in Griffith Park, the park home to the Griffith Observatory, the LA Zoo, and the Hollywood sign (Gammon 2022). Sometime before that, P-22, whose name comes from a National Park Service study of pumas in Southern California, had made a remarkable journey from the nearby Santa Monica mountains that required crossing multiple major LA-area freeways. For the next ten years, P-22 lived in the 4,310-acre urban park and periodically wandered by houses in abutting neighborhoods like Los Feliz. He became an iconic symbol of LA's surprising capacity to support nature in such a majestic, wild form. That capacity had its limits, though. Eventually

P-22, whose home in Griffith Park was substantially smaller than the typical range of a mountain lion in the wild, was captured by wildlife officials for a health checkup (Martinez 2022). They concluded the puma had sustained injuries from being hit by a car and had other health issues, including a skin parasite likely contracted from contact with domestic cats. In light of this damning health assessment, P-22 was euthanized (Queally and Nelson 2022).

The big cat's death was a sad end to his story that prompted an outpouring of tributes. However, it is worth noting that P-22 nearly met his demise years earlier, when he was treated for a bad case of mange in 2014 (Kuykendall 2014). Biologists linked this condition to poisoning from rodenticides that were found in P-22's system. Ultimately, the puma made a full recovery from this condition and roamed Griffith Park for another eight years. Other mountain lions, though, were not so lucky. In fact, between 2012 and 2019, several other named mountain lions from the National Park Service study died from either confirmed or suspected exposure to rat poison. In 2015, P-34 was found dead on a trail in Point Mugu State Park, north of Malibu, and biologists confirmed that she had been exposed to anticoagulant poisons used as rodenticides (Kuykendall 2015). A similar scenario unfolded with three-year-old P-47 in 2019, who had no visible wounds when he died but was also found to have ingested anticoagulants (Bloom 2019). P-54 died after being struck by an automobile in 2022, but her death brought still more troubling news on the rat poison front: both she and the unborn cubs she was carrying tested positive for various rodenticides (Solis 2022a). Besides the City Hall rat infestation, the plight of mountain lions in the National Park Service study is probably the most widely covered story related to rats and rat infestation in the *Los Angeles Times*.

The purpose of this brief detour into the world of mountain lions and the wild edge of "ecotone" where LA's human-built infrastructure overlaps with predominantly nonhuman ecologies is to illustrate both the practical difficulties of reconciling nature with urban environments and the nuanced symbolic meaning embedded in that pursuit. P-22 became a charismatic symbol of coexistence with nature, a fulfillment of an urban ideal where nature is folded into the fabric of urban life and we are all the better for it. This happened despite the potential danger P-22 always posed as an apex predator roaming a city park.¹ In fact, the puma's misadventures occasionally served as a reminder of that danger. Once, he leaped over the fence of the Los Angeles Zoo to prey on one of its koalas (Serna and Branson-Potts 2016). The wellness check that led to his euthanasia was prompted in part by an incident where he killed a chihuahua and attacked other dogs, behavior that officials interpreted as a sign of distress (Solis 2022b). This is all to say that it was not necessarily inevitable that P-22 and his fellow mountain lions would be revered more than they are feared. Nonetheless, they are generally viewed as a form of "good" nature that should be celebrated and protected.

The Indoors/Outdoors Divide

Cities are not only tasked with cultivating good nature, though. Properly welcoming nature into urban areas requires keeping "bad" nature like rats at bay as well. The controversy surrounding the poisoning of mountain lions who ingest anticoagulant-laced rats or other prey further up the food chain shows how these two goals—welcoming good nature and relegating bad nature—are sometimes at odds with each other. This dynamic was also evident in the city's decision to remove landscaping from the exterior of the Civic Center as part of its response to the rat problem inside City Hall.

What guides the management of urban nature, in both its good and bad forms, given these tensions? The City Hall rat infestation is instructive on this question because rats are simultaneously a public health problem and an enduring fact of city life. They are nonhuman animals that have become fixtures in urban environments, yet they are almost always considered a scourge to be eliminated or at least kept at bay, rather than a positive feature of a healthy city. On one hand, rats

---

1   Multiple incidents of California joggers crossing paths with mountain lions where hiking trails overlap with their habitat have made headlines over the years. A Sacramento-area woman was killed in 1994 (*Los Angeles Times* 1994). In 2022, a viral video shot by a jogger near Pyramid Lake documented his close call with an onrushing puma that retreated after he roared back at it. Instances like these inspired the opening scene of a 2005 episode of the hit HBO drama *Six Feet Under*.

are evidence of a point that many urban studies scholars have long emphasized: that cities are multispecies ecologies and thus that the boundary between nature and society is ultimately an imagined one. But while nonhuman animals make this urban ecology their home, they may become "problem" animals that transgress human-specific spaces (Jerolmack 2008). This makes rats a form of what I refer to as "bad nature": elements of the nonhuman world that are woven into the fabric of cities yet detract from the aspirational goals of urban management as opposed to furthering them. Beyond pests like rats, who compromise city life primarily by posing public health threats as vectors for infectious disease, other forms of bad nature in LA include the floodwaters that the area's rivers occasionally produce and the firestorms that inevitably erupt in the woodland ecology around the city's mountainous perimeter. By contrast, good nature, which often comes in the form of greening efforts, is crucial for crafting a desirable, quintessentially urban environment. Street trees and parks, for instance, aid in improving quality of life for residents by helping cool neighborhoods and improve their air quality. Cities, then, must embrace these good forms of nature while mitigating or eliminating the negative effects of bad nature.[2]

In a practical sense, rats are not leaving Los Angeles anytime soon. While the city devotes many resources to rat control, no one has any illusions about the possibility of eradicating rats completely. This sets urban rat control apart from the other forms explored in this book, where a rat-free zone is either being protected or possibly established. Rather, the goal in LA is to keep rats from breaching the indoors, where possible. This threshold, what I refer to as the "indoors/outdoors divide," is a key spatial logic underpinning how urban centers like Los Angeles manage nature. In other words, the goal of welcoming good forms of nature into outdoor areas of a city

---

2   Of course, the distinction between good and bad nature is not a neat binary separation, any more than the nature/society divide is a stark, empirically observable boundary. Neither are the two categories static. As attested to by the reverence for a dangerous predator in Griffith Park discussed above, the categorization of the nonhuman world along these lines is a meaning-making process that is accordingly fluid.

while preventing bad nature like rats from breaching the indoors is a way of operationalizing a preferred imagination of urban nature.

Ultimately, a neat, binary division between indoors and outdoors is as illusory in its material reality as the nature/society divide is. For instance, some human geographers have explicitly called for greater attention to indoor environments as "active political-ecological spaces" that are inseparable from the processes of the outdoor world (Biehler and Simon 2010). Dawn Biehler (2013) in particular notes the capacity of urban pests, including rats, to disrupt taken-for-granted notions of "public" versus "private" by permeating the indoors. However, it is this same hybridity that makes the indoors/outdoors divide important from a perspective of shared meaning. Though it may contain inherent contradictions, the indoors/outdoors divide emerges as a central symbolic axis guiding city officials' attempts to manage both good and bad nature.

The project of maintaining the indoors/outdoors divide through rat control, then, has both material and symbolic components. Granting that the threshold is ultimately permeable and that indoor spaces are not completely cut off from the broader ecologies of urban environments, LA officials would nonetheless like to ensure that rats do not physically breach the walls of the Civic Center. Equally, if not more important, however, is the symbolic work of guarding public faith in the indoors/outdoors divide. Making good on Herb Wesson's statement that city workers should not have to "worry about rodents" means giving those workers confidence that their work areas are comfortably siloed from the outside world. As the following sections explore, the symbolic importance of this work is derived from the uniquely potent cultural meaning that rats themselves carry.

TYPHUS AS BAD NATURE

For the average office worker, the sight of a rat in their workspace would be an unwelcome development regardless of the level of real, physical threat the rat posed. When the 2019 City Hall rat infestation happened, though, the rats in question did in fact pose a legitimate health risk. The rat problem coincided with a cluster of typhus cases in downtown Los Angeles, and eventually one City Hall employee would sue the city after contracting the disease, claiming that the dangerous conditions of a rat-infested City Hall were to blame (Zahniser 2019). This outbreak of typhus, however, was just the most recent of several outbreaks across the LA area during the past decade. Understanding this phenomenon requires examining such outbreaks not only as discrete events, but as episodes in the larger context of greater LA's ever-evolving relationship with the natural world. An urban outbreak of a vector-borne illness can be understood as an "entanglement" of different forms of life, an ecological balance of humans, fleas, rats, the typhus bacteria, and more, in the same way that Alex Nading (2014) describes the entanglement of people, mosquitoes, and dengue virus in Nicaragua. Moreover, the shifting balance of these ecological relationships in response to social and environmental factors can precipitate outbreaks of diseases like typhus. Viewed in this way, typhus outbreaks are problems of the management of good and bad urban nature.

One notable outbreak happened in 2015 in Pomona, a city in California's Inland Empire roughly thirty miles from downtown LA. Along with other outbreaks of the disease, this cluster of cases was linked to infected opossums who served as vectors (Wekesa et al. 2016). In a quotation provided for a *Los Angeles Times* article, Stuart Cohen, an infectious disease expert at UC Davis, explained the phenomenon in terms of increased interaction between humans and wildlife: "As some of these suburban areas encroach upon some of these habitats for opossums and other rodents, the chances of getting flea bites and therefore picking up the infection increases" (Karlamangla 2019).

While this explanation highlights how typhus and other diseases like it are problems of urban nature and the struggle to manage it, the spread of suburbia is just one aspect of a much larger socio-environmental history of typhus in Southern California. In fact, it would be somewhat misleading to view the development of new neighborhoods as having encroached on the habitat of the opossums that became key nodes in the transmission of typhus in several outbreaks. Opossums are not native to California but were introduced to the state during the late nineteenth and early twentieth centuries (Krause and Krause 2006). On multiple occasions, transplants from the South and the East Coast brought opossums to California either to serve as a source of food or to be raised for their fur, which was then used for garments. The wild population of the animals was established when some of them escaped their enclosures or, in one case, were voluntarily released to the wild by a fur farmer frustrated with the failure of this enterprise (Krause and Krause 2006). Before they became agents of bad nature by spurring outbreaks of disease, then, opossums occupied very different roles as livestock.

It is also important to note that the role of opossums in typhus outbreaks was facilitated by their contact with domestic cats and dogs, who in turn exposed humans to the fleas that carry the disease. This aspect of the epidemiological story further demonstrates how typhus outbreaks are incidents in the ongoing negotiation of good and bad forms of urban nature. Experts have pointed to a specific policy decision known as the Hayden Act (named for Chicago Seven activist turned California state senator Tom Hayden) as helping create the conditions for outbreaks (Wekesa et al. 2016). Passed in 1998, this law established new regulations related to animal control with respect to companion animals like cats and dogs. Specifically, it changed the minimum period of time that animal shelters and pounds must keep stray cats and dogs before euthanizing them from seventy-two hours to six days. To further the goal of reducing or even eliminating the euthanasia of stray animals, some counties adopted "trap-neuter-return" policies, under which stray cats would be caught, neutered, and then re-released where they were found. In a conference paper from the *Proceedings and Papers of the Mosquito and Vector Control Association of California*, Wekesa et al. (2016) argue that these measures caused the population of feral cats in Southern California cities to balloon and thus increased the contact between cats and typhus-carrying opossums.

In these authors' words, the Hayden Act and other policies aimed at reducing euthanasia in animal shelters "shifted the mission of animal control agencies from being public health to one of animal welfare" (Wekesa et al. 2016). The tension captured by this distinction and the pair of undesirable outcomes in the balance (mass euthanasia of animals on one hand and outbreaks of potentially deadly diseases like typus on the other) are indicative of the stakes of the management of urban nature. Municipalities and county governments must balance the protection of good nature, including charismatic domesticated animals like cats and dogs, against the goal of maintaining safe, healthy urban environments free from dangers to residents like those posed by infectious diseases. The downtown LA typhus outbreak and the City Hall rat infestation, though, show how the symbolic work that goes into managing this tension operates somewhat independently from the material work. It was this event that made typhus a visible public issue in Los Angeles. In fact, after the reports of the rat problem, the *LA Times* ran an article titled "Long before City Hall Rats, L.A. Has Struggled with the Rise of Typhus," which mentioned prior outbreaks like the 2015 Pomona one and thus informed the paper's readership that typhus was not a new phenomenon in the city (Karlamangla 2019).

Why did rats become such a potent symbol of the threat of disease when opossums and feral cats before them did not? Part of the answer to this question, of course, relates to the dramatic and inherently symbolic location of City Hall as a site of rat infestation. But it also points to rats' unique power to capture the public imagination as a symbolic boogeyman, even beyond their specific roles in the transmission of disease. As the following section explores, this is partly due to the notion of an "infestation" itself, which requires a transgression of the indoors/outdoors divide.

THE OUTSIZED SYMBOLIC POWER OF THE RAT

Compared to the two other case studies included in this book, in which rat control or eradication is the undertaking of a particular organizational body for the most part, Los Angeles's approach to its City Hall rat problem has been diffuse, touching the jurisdictions of a variety of different bureaucratic departments and outside agencies. As one of my interviewees, Deborah, an employee in the city's Department of General Services (GSD), described the scope of the city's response, "a number of different agencies, LAPD, L.A. Sanitation, the Mayor's Office, Personnel, Recreation Parks, City Attorney, Bureau of Streets Services, GSD, Housing, all of us were involved. We all had a piece of it and we just did our job." The city's Building Maintenance Division does have a contract with a private exterminator who carries out the parts of the job that most specifically fall under the banner of "rat control," but the way most of my interviewees from the City described things, the actual baiting and poisoning of the animals seems almost like an afterthought. The wide-reaching scope of the rat infestation and the number of different organizational bodies with no obvious ties to animal control involved in it hints at the sprawling significance of rats in the cultural imagination.

The city does have officials specifically concerned with rat control, including Mas Dojiri of the Department of Sanitation. However, Dojiri's position (chief scientist) is much wider in scope. In fact, when I asked him to describe his position when we first spoke, he listed several different duties, ranging from monitoring pollution in LA's various bodies of water, to overseeing the proper disposal of industrial waste, before realizing that he had omitted his duties as a co-leader for the "rat abatement strategy for the entire city of LA," which was the reason we were talking in the first place. This momentary oversight underscores the place that rat control occupies within LA's broader management goals; controlling the urban rat population is one piece of a broader public health mission aimed at keeping the city clean and safe. In fact, when he went on to describe his "rat abatement"[3] work, Dojiri presented the practice of rat control as less a goal unto itself than a means to a different end. The beginning of this description bore little connection to rats, period. He explained that when he began his current job, one of the things he was tasked with was managing an emerging epidemiological crisis, not from typhus, but from another disease: "Right after I got promoted to this position, we ended up having a hepatitis A . . . a little bit of a problem in the Skid Row area. . . . So, as chief scientist, I was in charge of making sure it didn't spread and that everybody was vaccinated who worked in the Skid Row area, etc. So, I worked with the [Los Angeles] Health

---

3   As a descriptor for Dojiri's duties, the term "abatement" is an indication of LA's relationship with rats and rat control relative to the other sites in this book. Rats are to be kept at bay, not eliminated completely, in the city center.

Department on this and then we ended up training Orange County Public Health and San Diego Public Health on our cleaning protocols so that they could kind of follow suit."

Hepatitis A, an infection of the liver, is highly contagious, but not in a way that has much to do with rats. The virus that causes the disease is found in the stool and blood of infected people, and individuals become infected when they ingest even a very small amount of it. As Dojiri put it, hepatitis A is transmitted via "fecal oral pathway," which means the most important areas of concern when it comes to prevention are food and water contamination. However, he explained, the hepatitis A problem then gave way to a typhus outbreak in downtown LA. Typhus is a bacterial infection that causes fever and other flu-like symptoms. While the vast majority of people recover quickly when treated for the disease early after contracting it, typhus can be fatal without treatment. In terms of transmission, typhus differs from hepatitis A in that it is not spread from person to person. This, Dojiri explained, necessitated a different kind of approach: "Typhus has a vector. It's not a fecal oral pathway. It's a vector pathway. . . . The bacterium infects fleas, which are on the rats. And then fleas end up biting . . . mammals. And then we end up getting typhus with this bacterium. So, we needed to control the rats."

As Dojiri describes it, controlling the rat population was a necessary measure for solving a problem that extended far beyond rats. In fact, he notes that, in terms of how people might actually come down with the disease, typhus is more closely linked to fleas than to rats. Nonetheless, the actionable solution was to indirectly control the population of fleas, the true disease vector, by controlling their hosts, the rats. As discussed above, typhus outbreaks are outcomes of a complex ecological system that are influenced by human policy decisions and other factors. Visualizing the depth of this interconnection can be difficult. In terms of both our lay conceptualizations of the typhus outbreak and, as Dojiri notes, the practical strategies for addressing it, the rat serves as a proxy for the totality of these entangled relationships. They both symbolically embody the threat of typhus and represent a targetable node to stop its transmission. It was notable, though, when asked about how he got involved in the rat abatement program, that Dr. Dojiri began not with typhus, but with hepatitis A, a disease without the same material connections to rats. Nonetheless, for him it was a relevant piece in the narrative account of his rat abatement duties. This is a good indication of how rats' symbolic associations with disease at large transcend even their actual role as vectors.

This looming symbolic capacity of rats was further demonstrated by the specific case of City Hall's rat problem. When our conversation turned to this topic, I was surprised by Dr. Dojiri's characterization of the issue:

MAS DOJIRI: It wasn't "rats" in City Hall. It was one rat.
AUTHOR: Really? OK.
MAS DOJIRI: I think it was one rat, if I'm not mistaken. And they did a bunch of traps, they set up a bunch of traps and they didn't collect any. And they set up traps for the fleas and they didn't get any fleas in City Hall either.

To hear this from an expert like Dojiri was intriguing, to say the least, given that my first introduction to this case was a February 7, 2019, headline in the *LA Times* that read "L.A. City Hall, Overrun with Rats, Might Remove All Carpets amid Typhus Fears." Beyond this news coverage, public concern over rats in City Hall was enough to compel Los Angeles to initiate the multifaceted response described by Deborah above, and to fill the agendas of multiple city council meetings with reports from expert employees on the problem and the steps taken to address it. The day after this headline ran in the *LA Times*, LA City Council President Herb Wesson Jr. introduced a motion that called for a number of different steps to address the problem and seemingly treated it as an urgent issue. The motion states that "there has been a noticeable increase in the volume of rodents in the area and within City buildings" and even makes reference to a flea problem in Wesson's own office, which it attributes to "the rodent issue." Upon introducing the motion during the council's session, Wesson emphasized the need to "make sure that we, as quickly as possible, begin to make sure that our constituents and all of our staff feel safe." In short, both the media reporting on the problem and official communications from the city government regarding it treated it as an urgent and verifiable problem.

When I expressed my surprise at this disparity, Dojiri indicated that he thought the circumstances of the issue distorted the actual scope of the problem, remarking: "The problem is it was televised. I think it was in a councilman's office and they ended up having some video on it, and so that video pretty much went viral." He went on to note the claim made by a city attorney that they contracted typhus while at work in City Hall, suggesting that this provided a backdrop of public health concern that contributed to the notion of a rat infestation causing public alarm. Dojiri was not the only one of my interviewees with close connections to the issue that explained things this way. Ryan, an employee of the private pest control company contracted by the city to conduct inspections and other pest control services, also felt the problem had been exaggerated. He dismissed the outcry over rats, remarking simply, "Media. It's the media," before going on to note, similarly to Dojiri, that "anytime you have something like that, like a claim of typhus from a city employee . . . it's going to get a lot of attention. So that really escalated things." Both Ryan and Dojiri see the worry over rats as primarily symptomatic of a broader atmosphere of fear related to typhus. Ryan, for his part, added that the physical evidence of rats in City Hall was minimal, as his company had "gone through a complete inspection throughout City Hall. . . . We've done insect monitoring traps for fleas and rodents. But [the reports of infestation] just got a lot of attention."

For comparison, the *LA Times* article where I first encountered the story of the infestation seems to suggest more concrete evidence than either Ryan or Dr. Dojiri implied there was. The piece begins like this: "At a Halloween celebration at City Hall last year, a rat gnawed through a pumpkin put out for decoration. In another incident, city workers found a dead rodent decomposing in an office ceiling. And then there were the rat droppings spotted on at least two different floors of the downtown building." Upon returning to the article after conducting these interviews, though, I did notice how, after this initial paragraph, the descriptions of the rat problem are much less immediate. For instance, it returns to the Halloween pumpkin incident, describing how employees "discovered that a pumpkin had been 'gnawed out' by an animal," but does not indicate that anyone caught a rat in the act. It then goes on to describe another, more indirect rat encounter, wherein someone "saw the tail of a rat as it scurried behind her couch in her office."

As a sociologist and a cultural analyst, my intention in noting these discrepancies is not to attempt to adjudicate the validity of either side's claims. Rather, the very uncertainty over the scope of the rat problem suggested by these wildly varying accounts is itself indicative of this story's capacity to capture the public imagination. Whether material reality, cultural myth, or both, the idea of a rat-infested City Hall is uniquely symbolically compelling, given rats' larger potent associations with disease. Understandably, people like Dojiri and Ryan, whose job it is to control the rat population, are focused primarily on the verifiable, material presence of rats. The descriptions in the *LA Times*, though, in which rats always remain just out of sight but leave behind unnerving traces like bite marks on a pumpkin, evoke a more generalized sense of discomfort and looming potential danger that transcends the animals themselves. Whatever the true extent of the infestation, the rat nonetheless stands in as a symbolic icon representing this uncomfortable feeling that something is not right. In the months after the initial reports of infestation, the rat took on an even broader symbolic association with dysfunction and disarray; a political demonstration installed a giant inflatable rat outside City Hall, which functioned as a double entendre for the animals themselves said to have infiltrated the building's physical walls and for corruption in city government.

Viewed in these terms, as a cultural phenomenon, City Hall's rat infestation was an event through which the public made sense of a complex urban ecology and its intersections with the surrounding social dynamics. Rats are cultural metonyms for disease, and the notion of a rat infestation is a perfect embodiment of a larger health issue like the typhus outbreak and the struggles to contain it. City Hall, of course, is itself a metonym for local governmental power and authority, which further heightens the symbolic power of the narrative of infestation. In his remarks at the city council meeting where the abovementioned motion was introduced, councilman Wesson makes clear the symbolic associations with disease that rats carry, and in doing so evokes the ominous atmosphere these associations can generate amid reports of infestation. "There has been a lot of speculation [related] to typhus," he remarked, before continuing that "there has been a lot of speculation [related] to the types of *uninvited guests* that we might have in this building and throughout the Civic Center complex" (emphasis added).

By functioning as "uninvited guests," rats perform symbolic work that crystallizes spatial boundaries in urban areas like LA. Whether the rat infestation was the nonissue that some experts like Dojiri regarded it as or was closer to the situation described in the *LA Times*, the idea itself of rats breaching City Hall inspired both deep concern and a significant mobilization of personnel and resources to re-exert control over the ecological conditions around the building complex. Of course, the urban ecology that rats stand in for contains more than just animals, city landscaping, and pathogenic bacteria. To that effect, the following chapter examines how the maintenance of the indoors/outdoors divide is inextricably connected to broader issues of power and inequality in LA.

THE ENDURING MEANING OF SPATIAL BOUNDARIES

The City Hall rat issue in Los Angeles sheds light on the material and symbolic dynamics of nature in urban environments. Rats are an omnipresent embodiment of "bad nature" that cities must manage. The goal of keeping rats at bay and preventing the public health problems they may cause is often in tension with the goal of welcoming good nature into city centers. The indoors/outdoors divide, I argue, is a key spatial logic guiding the approaches city administrations take to reconciling these tandem goals. Incorporating good nature into urban centers must be done in a way that prevents dangerous forms of bad nature from breaching the indoors. Occasionally, of course, maintaining what is ultimately an illusory boundary becomes a material impossibility. LA officials have made various compromises in different directions, favoring either good or bad nature, in such circumstances. These include removing plant life around the Civic Center that might compromise rat control efforts or tolerating a charismatic mountain lion in Griffith Park despite potential danger to people and domestic pets.

The notion of the indoors/outdoors divide itself is an attempt to rethink the notion of symbolic binaries like the "nature/society divide." Urban studies scholars have historically sought to critique and deconstruct such notions while bringing attention to how (1) cities support more-than-human ecologies; and (2) cities are inescapably tethered to socio-environmental processes that extend well beyond their supposed limits. This chapter's analysis does not contradict this core goal or the importance of considering cities as dynamically interconnected with the nonhuman world. Nonetheless, I show that, however fluid the boundary between society and nature is in material terms, the notion of a binary divide remains an important distinction invested with potent cultural meaning that underlies much urban policy and management. The "indoors/outdoors" divide is an important spatial logic in this arena that is fundamentally connected to the notion of a boundary between human society and nonhuman nature.

What we stand to gain by acknowledging the enduring symbolic importance of these divides, permeable as they may actually be, is a more nuanced understanding of the cultural meanings of nature. Specifically, the saga of LA's City Hall rats prompts us to give greater attention to the role of "bad" nature in environmental problems and the solutions to them that we might envision. I have argued that part of the aspirational work of crafting a good urban life is not only propagating "good" nature but relegating "bad" nature to an acceptable distance. The same principle applies to the task of imagining positive environmental futures writ large. Traditionally, environmentalism, and especially environmental conservation, has regarded nature as something to be protected or saved. This is true for the current challenges posed by climate change, where grappling with its existential threat means preserving the livability of the planet and protecting nature in the broadest possible sense. In chapters 6 and 7, which focus on rat control in environmental conservation, the stakes of good and bad nature in this sense will become even clearer.

✕ FIVE ✕

BULKY ITEMS

The "Rat Problem" and the "Homeless Problem" in Downtown LA

On February 8, 2019, Los Angeles's city council convened for a meeting that included an agenda item regarding the reported rat infestation in City Hall. This included reports delivered by representatives from several departments within the city government on the steps they had taken to that point to investigate and address the rat problem. A representative from the Department of General Services (GSD) reported, for instance, that the city's pest control contractor had "[walked] the entire perimeter of City Hall" and recommended measures the city might take to reduce the chances that rats might breach the government building's walls. The same contractor, GSD reported, had also returned that day to begin a thorough inspection of the interior of the building, during which they would "walk every floor of City Hall" to identify any existing issues. Throughout this meeting, speakers floated various potential actions the city could consider, including regularly shampooing or even removing all the carpets in the building and removing all plant life both within offices and in planters around the exterior.

Most of the speakers seemed to share a similar attitude about the topic, namely that it was a serious issue that the City ought to act decisively on, and that the City was primed to do just that, given the resources being marshaled for this purpose. For some of those watching and listening, though, there was an elephant in the room during this conversation, and that elephant was the situation just outside. One speaker, Councilman Joe Buscaino, struck a very different tone when he spoke, laying blame at the city government's feet for precipitating the rodent issue and the coinciding typhus outbreak by ignoring what he saw as its root cause:

> I, like many of you, have been very frustrated with the issue of the typhus outbreak in and around downtown Los Angeles. And I believe that this typhus outbreak is consistent with the Mitchell Injunction. This body two years ago approved an ordinance that is safe, that is legal. It allows individuals—our most vulnerable individuals—to have 60 gallons of storage. And yet today that injunction is prohibiting our outreach workers to get to these most vulnerable homeless populations in and around the downtown Los Angeles area. So rats are a symbol of this injunction. They are emblematic about how we lost control over the encampment issue, and if we can't protect the greatest symbol, our own City Hall, if we can't protect our own staff from this medieval disease, then we should all pack up and go home.

The "Mitchell Injunction" that Buscaino referred to is the outcome of a court case filed by several homeless individuals and advocacy organizations in 2016 against the City of Los Angeles (Anon 2016). The plaintiffs in this case charged that LAPD officers had been arresting unsheltered homeless individuals for minor, nonviolent offenses like sleeping on the sidewalk, despite the fact that these typically result in infractions rather than arrest. They also complained that these officers would then seize and destroy the personal belongings of the homeless residents they arrested, which sometimes included medications and other medical devices. In 2019, the city and the plaintiffs settled the case out of court. As a result, the city paid damages to the plaintiffs and agreed to significant policy changes surrounding the policing of sidewalk encampments (Chiland 2019). Specifically, the police had many new restrictions placed on their ability to seize the property of homeless individuals. The issue of public storage of personal belongings has become a particularly polarizing issue within the broader topic of homelessness in LA. In fact, in 2019 a group of business owners and residents in the Skid Row area of downtown LA formally objected to the settlement and moved to intervene in its enforcement.

With this context in mind, it becomes clear what Councilman Buscaino meant when he said the rats in City Hall were a "symbol" of the Mitchell Injunction. That statement draws a direct line connecting the issues of rats and typhus with the City's struggles with homelessness and housing insecurity. More specifically, Buscaino and others who share his interpretation believe that the settlement of the Mitchell case and the City's decision to place protections on homeless residents' right to keep and store personal belongings on the sidewalk has exacerbated the homelessness crisis and created the conditions for problems like typhus and rats. At the time of the city council meeting in early 2019, Buscaino's perspective was an exception within the broader discourse on the rat problem; few other speakers that day drew any explicit or implicit connections between the issues of sidewalk encampments and rats. However, months later, the *Los Angeles Times* reported that the pest control company contracted by the city, in their recommendations based on their inspection of City Hall's exterior, concluded that the encampments of unhoused residents had produced debris including food scraps and human waste that created the necessary conditions for the rat problem.

For those who share Buscaino's perspective, the rat problem in City Hall served as a symbolic referendum on the city's policy on homelessness, and especially its decisions that limited its capacity to police sidewalk encampments and enforce "quality of life" laws. For others, the rats were chickens coming home to roost; the failure of the city government to provide widespread access to housing had left its most marginalized citizens to live among rats and bear the brunt of a typhus outbreak, and so rats breaching the walls of City Hall forced the powers that be to look these same conditions in the face. In either case, the rat infestation could not be disconnected from the issue of homelessness. Addressing it, as I will show, necessarily bled into the arena of homelessness policy.

While the recommendations of LA's pest control contractor established a material connection between homelessness and the threat of rat infestation, these two issues share important symbolic connections as well, uniting them under a broader politics of public space and the management of urban nature. The previous chapter explored how the rat infestation in City Hall is a case of the management of urban nature in its good and bad forms. An important symbolic and spatial axis guiding that management of nature is the indoors/outdoors divide. In these terms, the City Hall rat problem and the enduring presence of homeless encampments in Skid Row and around the Civic Center represent inverse transgressions of the indoors/outdoors divide. Rats, of course, bring bad nature and its potential hazards to the supposedly safe indoors. Sidewalk encampments, meanwhile, represent the domestic life of the indoors spilling out onto the outdoors. In the process, this subverts "good" nature in several ways. For vehemently anti-homeless residents, the presence of encampments in public parks, for instance, compromises their aesthetic value. Moreover, for unhoused residents themselves, exposure to the elements makes otherwise good nature like LA-area rivers dangerous to their health and the security of their belongings.

This chapter investigates these symbolic connections between homelessness and the City Hall rat problem, from the perspective of the city's management of both issues. I rely on a combination of computational text analysis, using the *Los Angeles Times* as a window into public discourse on homelessness, and ethnographic data from a ride-along with LA's Department of Sanitation as they conducted one of the weekly cleanings of the Civic Center that were introduced as a response to the rat problem. I argue that initiatives that touch the issues of both homelessness and rat control (1) represent city institutions endeavoring to protect LA's capacity to afford the possibility of an urban life deemed widely desirable; and (2) do so by managing the city's relationship to urban nature by protecting confidence in the indoors/outdoors divide. The management of the divide is steeped in issues of power and inequality, resulting in contestations over what the goals of this project are and what amounts to a transgression of the divide in the first place. The discursive connections drawn between homelessness and rats (both as inverse spatial transgressions and as interconnected pieces of a public health problem) further highlight these issues of power and inequality by lowering unhoused people to the status of vermin.

TRANSGRESSING THE DIVIDE

Managing the indoors/outdoors divide is the overarching goal that unites the issues of rat control and homelessness in Los Angeles under a single public health crisis to be mitigated. As the previous chapter discussed, the city's response to the rat infestation is part of the broader, ongoing effort to manage good and bad urban nature in terms of the spatial boundary represented by the indoors/outdoors divide. Importantly, both rats and homelessness are phenomena that remind us of the material ambiguity of this divide, despite how it shapes our experience of city life. By doing so, they require the maintenance of what should be a taken-for-granted distinction.

The result of rats' transgression of the divide is the looming sense of danger epitomized by the consternation over typhus in City Hall. Rats bring the hazards of bad nature to the indoors, shaking public faith in the boundary that insulates us from these harms. Homelessness and especially sidewalk encampments transgress the divide in an inverse way. Informal tent shelters and shopping carts full of personal belongings represent the indoors inappropriately spilling over the divide onto the outdoors. It is important to note that LA's homeless crisis is much larger than these visible transgressions. Most of the homeless population in LA falls into the category of "sheltered homeless," spending nights under the roof of a dedicated homeless shelter or other temporary accommodations. Moreover, still others might be deemed housing insecure, living in overcrowded apartments to make rent in LA's notoriously exorbitant housing market or otherwise perpetually on the cusp of losing their access to housing. This is all to say, the housing crisis in Los Angeles (and elsewhere) is a structural problem that extends well beyond the most visible forms of homelessness.

This highlights an important feature of the indoors/outdoors divide and the transgressions of it, namely that those transgressions themselves are subject to contested meaning. For those more inclined to consider the broader context of housing insecurity in LA, homeless encampments are symbolic of a larger social structural failure to provide widespread access to safe indoor space. In a 2021 op-ed for the *Washington Post*, sociologist Neil Gong crystallized this perspective. He noted that punitive policing of homelessness, including mass evictions of encampments, is unacceptable and unjust, but cautioned that, on their own, tolerant policies allowing such encampments to persist leave the larger issues of housing accessibility intact. In his words, such a system is "a Frankenstein's monster created by mating civil libertarianism with austerity" (Gong 2021). In short, while police sweeps ought to be denounced, moving the residents of these encampments indoors should be a priority. For others, the same encampments are simply eyesores and physical impediments, and a priority should be placed on removing them and their occupants or relocating them out of sight. For instance, the DTLA Alliance for Human Rights, the coalition of business owners and residents who sued to stop enforcement of the Mitchell Injunction, describe their legal efforts on their website as an attempt to "require the City and County to take responsibility for their legal obligations to maintain a safe and healthy environment for all" (DTLA Alliance for Human Rights n.d.) and also advocate for a "return to clean sidewalks." In a motion filed as part of their litigation effort, the organization describes the problem this way: "Encampments are preventing LA Alliance members from reasonably enjoying the use of their property." Both of these perspectives are grounded in the notion that the indoors/outdoors divide is an important, meaningful spatial distinction for a functional urban environment, but the broader goals associated with maintaining that division as well as the methods for doing so are contested.

Relatedly, not all transgressions of the indoors/outdoors divide are created equal. Sidewalk encampments and City Hall rats prompt consternation over the divide and demand some sort of intervention, but other instances where the boundary between indoors and outdoors is blurred go largely unnoticed. Rats themselves frequently transcend the indoors/outdoors divide in ways that draw much less attention, including by living in nominally "indoors" spaces like parking garages. Meanwhile, as noted, most people experiencing homelessness in LA spend many of their nights inside shelters or hotels. Still other apparent transgressions of the divide are much more acceptable. An outdoor spin class in Venice Beach, for example, is an intentional breach of indoors/outdoors that is not only tolerated but may also be a lucrative business model. A potted fern or ficus brings nature indoors in a conventional, acceptable way. These counterexamples highlight how the indoors/outdoors divide and enforcement of it are deeply rooted in power and inequality. While Angelenos with power and capital can freely blur the lines between indoors and outdoors, the visibility of both the City Hall rat infestation and makeshift encampments reinforces the symbolic potency of the boundary they transgress. To that point, excited customers have historically set up camp on the sidewalk in American cities to await much-anticipated product releases from retailers like Apple, and this phenomenon has scarcely inspired the strong emotional reactions that similar-looking homeless encampments do.

Another aspect of these transgressions that reveals the dynamics of power and inequality beneath them is their connection to notions of sanitation and health. The discursive connections between rats and homelessness in public discourse lie in the topics of epidemiology and public health (a point most notably captured by the typhus outbreak in downtown LA that provided the backdrop for the public outcry regarding the City Hall rat infestation). Importantly, the typhus outbreak predominantly afflicted the homeless population but gained public scrutiny following the lawsuit filed by a city employee who contracted the disease, indicating how such public health "crises" are socially constructed within the context of existing social inequalities.

Put simply, the management of the indoors/outdoors divide is inherently an expression of power rooted in the fundamental question of who (and what) has access to public (as well as private) space. Other scholarship has examined how urban nature and nonhuman animals have acted as an arena of contestation over competing visions of urban life that reflect deep-seated urban inequalities along lines of race and class (Aptekar 2015; Loughran 2017; Mayorga-Gallo 2018). This chapter's topic sheds light on how this plays out through various forms of governmental administration: LA's mitigation of a typhus outbreak represents an institutional attempt to protect Angelenos' quality of life, but it also represents the city's failure to provide widely available housing. In this sense, the spatial management of the indoors/outdoors divide exemplifies the tensions between the sanitation and public health prerogatives of cities and urban inequalities; marginalized communities bear the brunt of both infectious disease outbreaks and the often punitive approaches to addressing them (Sewell 2020). While measures like regular cleanings of the exterior of the Civic Center manage the indoors/outdoors divide, they also serve to confine the dangers of "bad" nature to the unhoused without addressing the broader lack of affordable housing that creates the necessary conditions for spatial transgressions of the divide in the first place.

THE SPATIAL MANAGEMENT OF HOMELESSNESS

The demographics of homelessness in Los Angeles highlight the enduring inequalities of the city and the United States at large. Specifically, people of color make up a disproportionate percentage of persons experiencing homelessness in LA. Most strikingly, according to data from the 2020 Los Angeles Homeless Count, nearly half (49.1 percent) of the homeless LA population identified as Black/African American (Los Angeles Homeless Services Authority 2020), compared to only 8.8 percent of the city's overall population (Census Bureau 2021). The racial demographics of LA's unhoused population are important context for discussions of the spatial management of homelessness. For one, the racial disparities of this population show how unequally distributed comfortable access to indoor, private space is. Policing the physical locations of the unsheltered homeless, meanwhile, is inherently an expression of power that dictates who has access to certain public spaces as well. The fact that the population of homeless encampments disproportionately come from already marginalized communities makes this fact even more stark. Compounding the issue is the fact that punitive approaches to homelessness come coupled with a state mental healthcare system characterized by "abdicated authority" (Barnard 2023) that results in the seriously mentally ill systematically slipping through the cracks and repeatedly winding up on the streets.

An ideal place to examine symbolic aspects of public space and the enduring issue of homelessness is the pages of the *Los Angeles Times*. The paper currently has a dedicated "Housing and Homelessness" section, and has long devoted considerable print to the issue of housing insecurity in LA. Just between 2015 and 2019, a time frame selected to capture the years leading up to the City Hall rat infestation, 1,304 articles contained the word "homeless" or "homelessness" in the article title. These search parameters are of course only a very conservative snapshot of the scope of the paper's coverage of this topic, as many stories that do not have these words in the title surely deal with homelessness in both substantive and peripheral ways. As a source of data, the *LA Times* provides just one editorial perspective on this issue, but it is largely seen as the paper of record for Southern California and is one of the most widely circulated newspapers in the country. As such, it provides a general reflection of local discourse on homelessness as well as playing an important role in shaping that discourse.

In what follows, I use a computational text analysis of the *LA Times*'s coverage of homelessness to identify themes that cohere in discussions of the issue as well as to better understand how the city approaches the issue from a spatial perspective. Using a combination of word embeddings, dimensionality reduction techniques, and clustering algorithms, I identify and visualize clusters of semantically similar words in the text data that are taken to correspond with recurring themes in the articles. These themes can then be interpretively analyzed based on the terms that comprise them and the articles that engage them most heavily. For the sake of readability, I am confining a more technical description of the procedure I employ to appendix B.

As a general overview, though, the combination of tools I use works as follows: word embeddings use one of several algorithms to generate from a large dataset of text a geometric model where each word is given a set of coordinates. These coordinates represent words' relatedness to each other; similar or related words tend to have coordinates that are close to each other. With dimensionality reduction techniques, a researcher can convert these coordinates into a format that can be plotted on *x*- and *y*-axes. Finally, clustering algorithms group points in this two-dimensional space into mutually exclusive clusters. Ultimately, this allows for a visualization of the landscape of homelessness discourse in the paper where each point in a scatterplot represents a word. Figure 5.1 displays this visualization of terms in the *LA Times*'s coverage of homelessness with shaded boundaries outlining the clusters identified in the analysis.

FIGURE 5.1 Visualization of cluster analysis depicting topical themes in the *Los Angeles Times*'s coverage of homelessness. Each point is a word, and their relative positions indicate their semantic relatedness to each other, as calculated from a word embeddings model. Clusters identified by Gaussian Mixture Modeling are outlined and labeled with descriptive titles.

The most central terms[1] in each cluster are labeled, and the larger labels in italics are descriptive titles for each cluster assigned based on interpretive analysis. Examining the articles that engage most with a given topical cluster provides further insight into the cluster's makeup and how it corresponds to patterns of discourse related to homelessness. These articles can be identified using word embeddings by comparing the coordinates of the words used in a given article with those of the words that comprise a cluster.

Abstract and Concrete Discussions of Homelessness

The clustering algorithm identified nine thematic clusters in the data. The cluster titled "Issue Overview" corresponds to general discussions of the state of homelessness in Los Angeles and contains words like "significant," "impact," "issue," and "finding." Articles that heavily engage this cluster provide descriptions of the scope of LA's housing crisis or report the findings of studies related to the problem. Some of the headlines of articles exemplifying this cluster are "As L.A.'s Homeless Crisis Worsens, No One Is in Charge. That Has to Change" (Lopez 2019a), and "U.N. Monitor Says L.A. Lags behind Other Cities in Attacking Homelessness" (Holland 2017). Another broad category that abuts "Issue Overview" in the visualization is "Policy Proposals." With words like "plan," "proposal," and "measure," this cluster reflects reporting on governmental and legislative attempts to address homelessness through public policy. Example headlines from this cluster are "A Fix for L.A.'s Homeless Crisis Isn't Cheap. Will Voters Go for $1.2 Billion in Borrowing?" (Smith 2016a) and "Los Angeles County Outlines Strategies to Reduce Homelessness" (Sewell 2016).

These two clusters, along with "Politicking," "Legal Actions," and "Taxes and Funding," relate to discussions of homelessness in a register that is somewhat abstracted from the immediate, on-the-ground reality of the issue. "Politicking," which includes words

---

[1] The centrality of each word to its respective cluster is calculated by computing its eigenvector within an adjacency matrix containing only the given cluster's terms. A more detailed description of this process is in appendix B.

like "lawmaker," "democratic," and "candidate," corresponds with discussions of how, as a public issue, homelessness might impact other outcomes like political races. Meanwhile, articles that heavily engage the "Legal Actions" cluster report on litigation related to homelessness, such as "Homeless Man Wins Harassment Settlement from San Diego Police" (McDonald 2017) and "Deputy Retaliated against Activist Who Protested Clearing of Homeless Encampment, Lawsuit Claims" (Wigglesworth 2019). Finally, the "Taxes and Funding" cluster contains words like "tax," "rate," "economy," and "spending," and corresponds with articles like "[Los Angeles Mayor Eric] Garcetti's Budget Would Spend More Money on Street

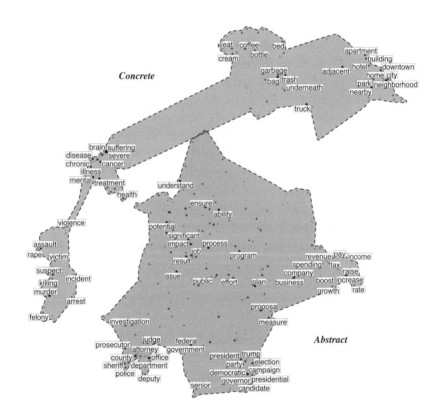

FIGURE 5.2 Second visualization of the cluster analysis with boundaries redrawn to encompass "abstract" and "concrete" engagements with homelessness.

Repairs and Homelessness" (Smith 2019) and "[HUD Secretary] Julian Castro Calls for Surge in Federal Spending to End Homelessness" (Finnegan 2019).

Compared to these first five topical clusters, the other four represent discussions of homelessness that are much more concrete and immediate. Whereas the clusters mentioned above engage the topic of homelessness as a public issue and discuss how it connects to other broad topics of interest like politics or the economy, the latter four clusters, "Public Health," "Violent Crime," "Life on the Streets," and "Politics of Space," focus on the lived realities of homelessness and the experiences and practical problems encountered head-on by those working to ameliorate it. Figure 5.2 displays the same visualization depicted in figure 5.1, but with lines redrawn to collapse the clusters into these broader "abstract" and "concrete" categories with respect to their engagement with homelessness. Partitioning the words into these new categories indicates that they occupy discrete spaces within the sematic environment of the word embeddings model, suggesting that they describe a meaningful distinction in the *LA Times*'s overall discussion of the topic. These more concrete topical clusters shed light on the spatial dynamics of homelessness and the city's management of it, as well as the cultural meaning that colors discourse about the issue.

Stigma and Moralization of Homelessness

Two of the concrete clusters, "Public Health" and "Violent Crime," indicate some of the negative associations that homelessness and unhoused citizens carry. The "Public Health" cluster contains words like "mental," "illness," "disease," "treatment," and "care." Many of the articles that engage this topic focus on the prevalence of mental health issues in homeless populations. Others discuss other health problems that homeless residents face due to the conditions of living on the street, a lack of reliable access to essential healthcare, and a widespread aversion to accepting the services on offer from nonprofit and governmental organizations. For instance, a pair of columns, one by a medical doctor and director of a nonprofit healthcare resource for LA's homeless and one by longtime *Los Angeles Times* columnist Steve Lopez,[2] both describe how Skid Row residents frequently decline treatment offered by outreach workers when it is critically needed and end up dying despite those providers' best efforts (Lopez 2017; Partovi 2019). Another article has the chilling headline "L.A. County's Homeless Population Is Growing—but Not as Fast as They're Dying." The article notes the impact of widespread substance abuse on the spiking mortality rate in the homeless population the piece reports on, but also states that "A stressful lifestyle, lack of healthy food and constant exposure to the weather all contribute to an early death" (Gorman and Rowan 2019).

Many of the articles that engage this topic, including the three cited above, are descriptions of the plight of unhoused citizens containing implicit or explicit pleas for policy changes or other measures aimed at easing this population's suffering. In other words, the presence of words like "severe," "illness," "cancer," and "disease" in the top right corner of figure 5.1 reflects the *LA Times*'s documentation of real hardship that befalls the homeless community, hardship that often becomes practically invisible when it afflicts those at the extreme margins of society. The "Public Health" cluster also indicates, though, that medical ailments (especially mental illness) are one of the themes most closely associated with homeless individuals in public discourse. This is revealing in light of the fact that concerns over an outbreak of typhus (one that most severely affected those on the street) was a key factor that made the City Hall rat infestation a major issue, and that homeless encampments were cited as a primary cause of the infestation. Whereas rats' symbolic role in discourse surrounding the typhus outbreak, as discussed in chapter 4, transcended their specific epidemiological role, the existing associations with homelessness and disease made them similarly apt as scapegoats.

Another one of the "concrete" clusters that sheds light on how

---

2   Lopez's coverage of homelessness-related topics for the *Los Angeles Times* has garnered national attention and inspired the film *The Soloist*, which stars Robert Downey Jr. as Lopez and centers on his relationship with homeless classical string musician Nathaniel Ayers.

many Angelenos might have been more likely to view homeless encampments as the cause of a public health crisis rather than its chief victims is "Violent Crime." With words like "assault," "victim," "suspect," "felony," and "shooting," this cluster reflects articles reporting on violence involving homeless citizens, either as perpetrators or as victims. The articles that most closely align with this cluster are roughly split in this regard, with headlines like "Homeless Man Suspected of Raping and Kidnapping a Woman from North Hollywood Parking Garage Is Arrested" and "Homeless Man Charged with Sexually Assaulting 4 Women in West L.A. and Santa Monica" accompanying others like "Man Who Killed 5 at a Long Beach Homeless Encampment Gets Life in Prison" and "Two Charged in Slaying of Homeless Man—the Third Person Killed This Year in Hollywood." These two genres of stories within this cluster reflect a similar duality to the one represented by homeless residents' status as both victims and perceived causes of disease outbreaks. The "Violent Crime" cluster suggests the homeless community is disproportionately vulnerable to being victimized by violence while also representing a source of danger to other Angelenos.

Both of these dual associations combine to generally link homelessness with violence. The frequent victimization of the homeless indicates how routine and mundane violence against them is, and the characterization of homeless citizens themselves as violent criminals implicitly explains or justifies that victimization. Important to note is that a subset of the articles that discuss incidents where homeless individuals were targets of violence relate to a police shooting of an unarmed homeless man in Venice Beach (Mather 2016a, 2016b; Mather, Chang, and Gerber 2018). Other articles report on police harassment of homeless and other punitive forms of policing. While the shooting in Venice is an extraordinary case (which resulted in a rare recommendation of criminal charges against the officer by the LAPD chief), this coverage generally indicates the routine and unremarkable nature of violence against the homeless community, including as a form of policing. In this context of both normalized violence against unhoused citizens alongside pervading associations that cast them as dangerous criminals themselves, both of which dilute their humanity, it is unsurprising that homeless encampments are viewed broadly as the source of social problems (like rat infestation and typhus).

The Materiality of Homelessness

While the "Public Health" and "Violent Crime" clusters shed light on the stigmatized cultural meanings ascribed to homelessness, the final two clusters are the most revealing regarding the spatial management of homeless encampments. First, the "Life on the Streets" cluster, which contains words like "bag," "trash," and "backpack," deals directly with the material culture of homelessness. Many of the articles represented by this cluster concern the controversies over public storage of personal items and the legal and legislative battles surrounding it. Beyond just reporting on these administrative outcomes, though, the *LA Times*'s coverage captured by this cluster provides a more intimate window into the realities of homeless life. One column headlined "The City Has a Growing Mountain of Possessions Confiscated from Homeless People" (Lopez 2019b) provides a glimpse into the massive storage facilities established to house the personal effects of the homeless. Another, "How the Homeless Live and What They Keep in L.A." (Holland 2015a), profiles six individuals living on the street and the possessions they retain, which include Bibles, cleaning supplies, and stuffed animals.

The storage of personal items and the controversies surrounding it highlight the transgression of the indoors/outdoors divide that living on the street represents. Mattresses laid upon surfaces of dirt beneath overpasses and shopping carts carrying cherished personal possessions represent the domesticity of indoor life spilling over the divide to the outdoors. While some *LA Times* coverage, like the profile of individual residents and their belongings, inspires empathy, the topic of so-called bulky items is a source of contention within the broader issue of homelessness. Some articles dwell on descriptions of these personal possessions, illustrating in detail the varieties of objects stored on the streets, their immensity, and the perceived impediments to city life their physical existence imposes:

Spangler told the citizens watchdog group that he has seen drug addicts loitering around the gym, stashing shopping carts filled to the brim with their personal belongings and blocking access to some of the equipment. (Reyes-Velarde 2018)

Many of the roughly 1,000 homeless people who live downtown regularly pack up their belongings before dawn on cleanup day to make way for crews, who remove about three tons of debris from the sidewalks each week. (Warth 2017)

The homeless camp was littered with heaps of broken furniture, disgorged computers, bicycle frames, televisions, disassembled motorcycles, pieces of exercise machines, rotting food, empty containers and half-buried clothes. (Smith 2016b)

One effect of these descriptions is the blurring of lines between people's possessions and garbage or even potentially toxic waste. Both discursively and materially, this stems from these possessions' status as a form of "matter out of place" in the sense in which Mary Douglas (1966) used the phrase. Douglas's concept describes how otherwise benign material objects become "dirt" when encountered outside their expected, appropriate contexts. Items like exercise equipment and electronics become taboo objects when they are removed from their proper place in the indoors/outdoors divide. What's more, objects at home on the indoors half of that divide become prone to physical deterioration, making them more unsightly, or with molding/contamination making them physically dangerous.

These belongings suffer from a divergence between materiality and cultural meaning, where their physical properties render them increasingly incapable of occupying an acceptable place in our meaning systems, inflected as those systems are with social inequalities. Cultural scholars like Domínguez Rubio (2020) and Terrence McDonnell (2016) have observed how the unrelenting ravages of time and physical processes rob material objects like these of their capacity to perform their previous symbolic duties. For personal belongings stored on the street, underpass, or riverbank, this physical change places them in a system of meaning wherein they represent a spatial transgression and a failure of the city to offer a desirable urban life. Moreover, as Robin Bartram (2021) argues, the social location of these possessions' owners shapes symbolic interpretations of them, compounding their associations with disarray and hazard.

The potency of the new meaning taken on by these items is evident in their enduring status as a subject of political and legal contention. This tension is exemplified by the controversy surrounding the previously mentioned Mitchell Injunction, which hindered city officials from confiscating unhoused residents' possessions without notifying them. For both unhoused advocates who celebrated this decision and other residents who opposed it, the issue of the right to publicly store one's belongings is indicative of a social structural failure of the city: for the former, it is a visible reminder of the inability of the community to provide basic needs to its most vulnerable citizens in the form of shelter and storage space, while for the latter, the belongings represent an impediment to businesses and other urban stakeholders.

Finally, the "Politics of Place" cluster contains the words "neighborhood," "park," "nearby," and "adjacent," among others, and represents discussions of the spatial location of homeless encampments. Like discussions of personal possessions cluttering the sidewalk, the related, core issue of where homeless citizens are to go is a source of much strife in LA. Some of the articles epitomized by this cluster focus on the ongoing challenges of building new shelters in the face of "NIMBYism" ("not in my backyard") and scarce real estate (Smith 2017). Others report on the expansion of homeless encampments into new parts of the city (Holland 2015b). A theme within these discussions is the visibility or seclusion of encampments within the city. These encampments are often discussed as existing either in the most visible of places, like downtown street corners outside government buildings, or the most marginal of places, specifically on the banks of LA and Orange County area rivers. This echoes the findings of other researchers who have emphasized how policy toward homelessness typically aims to either temporarily displace unhoused people to appease other residents who issue complaints (Herring 2019) or contain camps in marginal spaces away from public view

(Beckett and Herbert 2009; Herring 2014). Often, highly visible or extremely marginal camps were presented as an explicit dichotomy, suggesting that the forced eviction of encampments in one area necessarily leads to their emergence in the other:

> Carter has been trying to broker a plan and has warned officials that he doesn't want the homeless people displaced by the riverbed sweeps to end up at the Santa Ana Civic Center, which already is overwhelmed with homeless camps. (Money 2018)

> Another man living on the river who identified himself only as "Rabbit" said that people have flocked to the area because of the crackdown on the homeless in downtown San Diego. (J. E. Smith 2017)

> Then Orange County pushed hundreds out of an encampment along the Santa Ana River trail. Officials have also vowed to remove the tent city that has taken root at the Santa Ana Civic Center. But as the county and some of its cities take steps to push out homeless people, the problem of where they should go remains unsolved. (Do, Carcamo, and Serna 2018)

This dualism between city center and riverside encampments indicates the importance of the indoors/outdoors divide for the maintenance of good and bad urban nature and how the issue of homelessness itself is an explicit challenge to that goal. As the discourse around "bulky" personal items reveals, homeless encampments in highly visible public areas represent a spatial transgression wherein the domestic life of the indoors breaches the outdoors. One effect of short-term, non-systemic attempts at managing homelessness (i.e., encampment sweeps and evictions in the absence of systemic remedies to LA's affordable housing crisis) is to push these camps to marginal spaces in the domain of urban nature, like riverbanks. Given this, these short-term measures to clear public streets further compromise the aspirational project of maintaining "good" urban nature. The presence of camps on the sides of LA's rivers makes these spaces less capable of symbolizing good nature, and even transforms them into "bad" nature in the form of health hazards; several 2015 articles in my qualitative analysis detailed outreach efforts in riverside areas with encampments in advance of El Niño storms that threatened to destroy makeshift shelters and drown residents.

The threat of inclement weather to these riverside camps is an indication of how the indoors/outdoors divide shapes how nature might be good or bad vis-à-vis the goals of urban management. In general, rivers would fall into the good nature category, as visible signifiers of the nonhuman world located in appropriate spaces within the urban landscape. Homeless encampments bring the world of the indoors to the riverbank, though, allowing storms to breach this divide by causing flooding. It is worth noting that this particular example is an echo of Los Angeles's past when it comes to reconciling urban nature with livability. The city's rivers have long presented a public health problem in the form of flooding, and famously had to be tamed by an ambitious Army Corps of Engineers project that confined them with concrete and prevented similar breaches of the indoors by bad nature.

The four "concrete" clusters in this analysis reveal how the material reality of homelessness is a transgression of the indoors/outdoors divide, as well as how this transgression and the management of it are moralized (given enduring associations between homelessness and both violence and disease). This backdrop sheds light on homeless encampments' involvement in another such transgression, the City Hall rat infestation that they were cited as contributing to. The next section turns to the management of this issue and how addressing the rat problem necessarily bleeds into the contentious arena of homelessness policy.

MANAGING THE INDOORS/OUTDOORS DIVIDE

Clad in high-visibility orange vests on a Saturday morning, we sat in a pickup truck on Main Street, between First and Temple, periodically checking our phones and waiting to be cleared to begin the day's work. I was accompanying a crew of workers from Los Angeles's Department of Sanitation on a weekly cleaning of the Civic Center, the section of downtown LA home to the complex of buildings that makes up City Hall. These cleanings began as part of the City's multi-faceted approach to the City Hall rat infestation, to contribute to making the area a less inviting habitat for the rodents. It was already fifteen minutes past 7:00 a.m., the hour the cleaning was scheduled to begin, but we were not checking our phones for the time, or even for a text message or other outside communication giving us the go-ahead. Each of us was checking our phone's weather app. The forecast was for clear skies and a typical mid-February high to make most of the country outside Southern California jealous. At that moment, though, the temperature sat stubbornly at 49 degrees. That number would have to hit 50 degrees before we could begin, a protocol put in place for the protection of the city's unsheltered residents who would have to be roused for the cleaning to proceed.

 The area of Los Angeles most associated with its homeless population is Skid Row, which is located a few blocks away from the Civic Center, but in recent years the presence of tents and other makeshift shelters has grown increasingly prominent outside City Hall. While these Saturday cleanings were instituted in response to the rat infestation, in functional terms their main goal is to mitigate the physical effects that the phenomenon of people living outdoors has on this portion of the urban landscape. For the cleaning to proceed, anyone taking shelter in the area must at least temporarily relocate. In addition to the temperature, we were also waiting for representatives from the Los Angeles Homeless Services Authority (LAHSA), which is run by LA County and not by the City. LAHSA is supposed to make "first contact" with unsheltered residents and make them aware of public services available to them, but they had not shown up and no one in our group had been able to get in contact with them. Eventually, a police car drove by and announced over its loudspeaker that the area was scheduled for cleanup, and that anyone in the vicinity had fifteen minutes to move. I was told that this announcement cleared us to begin in lieu of LAHSA's arrival (which ultimately never happened). We waited a while longer as a man emerged from his blue tarp tent, which he collapsed and carried off down the street. A few others packed up to leave as well.

 These cleanings are a central piece of the city's attempt to exert control over the spatial dynamics of the Civic Center complex in response to the reports of rat infestation. Whereas rats embody illness, infection, and danger and mark the urban spaces they claim with these same associations, the work of the Sanitation Department reclaims these spaces and subverts those meanings. They do so by materially managing the landscape, altering its conditions so as to physically expel rats. The presence of homeless encampments in the area is a key factor for this city effort, in that they present specific obstacles for sanitation work and highlight the social politics of space within which this work implicitly takes place.

When the area has been cleared of people, the Sanitation team goes to work inspecting the visible remnants of the encampments. Most of the cleanup is simply clearing fairly benign debris like garbage strewn about the concrete. Some areas require more specialized attention, though, when they are deemed to contain hazardous materials. A small contingent of our group charged with managing these materials was dressed in stark white suits that offer more protection. For instance, early on, someone spotted a pile of trash from one of the vacated encampments with still-loaded hypodermic needles, presumably left behind by an intravenous drug user. The more common hazardous material, though, is urine and feces. We encountered these often enough that for the duration of the cleaning I consistently was reminded of a community swimming pool by the chlorine solution the crew sprays onto the concrete in these cases. This particular part of the job is noteworthy, as the presence of human excrement is one thing that experts specifically pointed to in their assessment of the area around City Hall as a factor for increasing the threat of rat infestation. Symbolically, it also highlights how unsheltered homelessness transgresses the indoors/outdoors divide and in the process compromises the acceptable role of urban nature. Human waste like this is not only "matter out of place," but it subverts the good nature of landscaping in planters and makes it toxic and unsightly. In terms of the cleaning's rat control goals, I could observe the important shift in the area's material conditions (from hospitable to rats to not) with my eyes closed, as a nauseating olfactory landscape gave way to the sterile smell of chlorine.

After making our way through the area immediately surrounding the entrance to the City Hall East building on Main Street, we descended a staircase to a courtyard situated below street level. There was not nearly as much garbage or other, potentially hazardous debris plainly visible there, but there were a few small piles of what looked like personal belongings, presumably left behind by people who vacated after the police made their announcement of the cleaning. There was also one man still sleeping on a bench. Miles, a supervisor of our crew, informed me that technically we should ask him to leave, as the area is supposed to be closed to the public during the cleaning, but since "he's not bothering anyone" we did not this time. On another bench across the courtyard sat a small knapsack. The protocol with items like this is to determine whether or not it is hazardous, and, if it is determined not to be, it is put in storage and replaced with a note indicating the facility where its owner can pick it up. Wearing plastic gloves, George, a member of the Sanitation crew, opened the bag up and immediately made his judgment. "Oh yeah," he says, "the second I opened it up I could smell the mold." This is one of the more common reasons belongings like this are determined to be hazardous. The bag would have to be thrown out.

As is clear from this cleaning protocol initiated after the reports of City Hall's rat infestation, the response to this problem is inherently tied to the city's policy on homelessness. The connections between these two concerns are both practical and symbolic. In general, what the Department of Sanitation must do for the ultimate goal of preventing rats from taking up in City Hall buildings is remove the physical effects of people living outside. A rat infestation like the one reported in City Hall can only be understood in terms of the urban ecology that provides the necessary conditions for it. In this case, a major factor in that ecology is the presence of human beings living on the street. Thus, the necessary maintenance to prevent rats from breaching the walls of these buildings involves regularly removing the remnants of the area's unhoused population, which include garbage, human waste, and (potentially hazardous) personal belongings. From a symbolic standpoint, removing these visible remnants and managing the material conditions of the urban ecology simultaneously marks the area as safe and ordered. When Councilman Herb Wesson urges that city employees need to feel safe when going to work, it is, in part, the work of the Sanitation Department to physically generate these associations for this building complex, by removing the atmosphere of looming danger that rats carry with them.

While downtown LA will likely support a robust population of rats as long as its human population remains, confining bad nature like rats to its proper place in the urban landscape means guarding the threshold between indoors and outdoors. Unlike Alberta and the Galápagos Islands, in Los Angeles, no one has any illusions about eradicating rats completely. Instead of a political border or the coastline of an island, then, the most important spatial boundary for the project of rat control here is this threshold between indoors and outdoors. Rather than envisioning a rat-free LA, the city is invested in a rat-free City Hall. Preserving the meaningful indoors/outdoors distinction means managing the elements of the urban landscape that transgress the popular conception of it. While rats are metonyms for disease and danger, attending to this divide means keeping the indoors safe from those associations. The looming sense of unease and the atmosphere of danger associated with the notion of a rat infestation is the result of a transgression of this boundary, where rats materially and symbolically bring the outside in.

The work of the Sanitation crews and of other departments like Parks and Recreation to transform the landscape immediately surrounding City Hall is an effort to seal these important boundaries and maintain a controllable, ordered environment that is an American urban ideal (Sennett 1970). This means materially removing the conditions in which rats thrive, but symbolically it involves removing the visible disorder of debris and scattered belongings and replacing the scent of human waste with that of chlorine spray. This visible disorder, on the other hand, is itself a representation of the frayed division between indoors and outdoors, not only because of the habitat it provides for rats, but because of the inverse spatial transgression of homelessness (as noted in the above analysis of *LA Times* discourse on public storage of bulky items). While the weekly relocation of homeless encampments for cleaning is necessary for the project of rat control, it also removes the visible remnants of the phenomenon of homelessness. Both of these—rat control and the maintenance that homelessness necessitates—are part of the overarching mandate to preserve faith in the spatial division between indoors and outdoors.

THE STAKES OF NATURE AND ITS LIMITS

The project of rat control in LA, as we have seen, is inescapably tied to the phenomenon of homelessness. In one sense, these phenomena are materially linked, where the social conditions that relegate a sizable population to living on the streets generate material conditions that make rat infestations probable. Symbolically, they represent inverse transgressions of the inside/outside divide, where rats bring the dangerous ills of the outside into indoor spaces and homelessness represents an inappropriate spillover of inside life onto the sidewalk. Accordingly, the rat control efforts of the city, including the weekly cleanings by the Sanitation Department, are invested in addressing both these issues simultaneously.

The intertwining of LA's "rat problem" and its "homelessness problem" further reveals how rat extermination has meaningful connections to social inequalities. As this urban case shows, the project of controlling and exterminating animal others is a symbolic expression of power and value. Securing the boundary of the indoors/outdoors divide against rats aims to preserve the safety and desirability of private spaces where access is unequally distributed. Moreover, the approaches that city organizations take to addressing issues like the spread of infectious diseases reflect underlying social inequities of urban life. The typhus outbreak in Los Angeles became a crisis not when it afflicted the homeless community, but only after a city employee contracted the disease. While, as explored in the previous chapter, rats have come to symbolize dirtiness in disease in ways that transcend their roles in epidemiological pathways, the implicit notion that unhoused citizens are doomed to suffer from the scourge of typhus symbolically lowers their own status accordingly; LA's homeless were considered to have helped "cause" the rat problem and the corresponding spread of typhus, though they were themselves vulnerable to that disease and forced to live among rats. Ultimately, of course, it is not hard to envision how a different approach to managing the indoors/outdoors divide might avoid this entrenchment of symbolic inequalities. This would involve structural action on the housing crisis that prioritizes first and foremost providing widely available access to safe housing, through a combination of robust public housing development and rent control.

The concept of the indoors/outdoors divide highlights the spatial aspect of rats' role in the negotiation of symbolic boundaries. The interspecies and social hierarchies touched on above are specifically linked to the spatial division between indoors and outdoors. This division, in turn, is a focal point for urban relationships to nature. Nature, on one hand, has become an increasingly central part of the creation of a positive urban experience, to the extent that parks, trees, and other landscaped plant life are overwhelmingly seen as providing social benefits and generally increasing city dwellers' quality of life. This imagination of urban nature, though, is dependent on the persistent notion of a nature/society divide, and a corresponding spatial order that allows these benefits of nature into cities while keeping at bay the deleterious effects that the nonhuman world may pose. The case of LA's City Hall rat problem illustrates this effectively. Rats are symbolically located on the boundary between nature and society, as nonhuman animals that thrive in urban settings. As metonyms for the dangers of dirtiness and disease, they are categorized as a form of bad nature that must be managed.

The boundary between nature and society is the most fundamental line that rats blur, and that rat control attempts to clarify. By turning to the role of rat control in environmental conservation practices, the next two chapters explore the widest implications of this meaningful boundary.

PART III

ECOLOGIES OF MEANING

✳ SIX ✳
ECOLOGIES OF MEANING IN ECUADOR'S GALÁPAGOS ISLANDS

Most people who visit Floreana Island depart from the Galápagos Islands' main population center, Puerto Ayora on Santa Cruz Island. The trip is two hours due south by boat, in my case a small vessel called the *Queen Astrid* that was near its maximum capacity carrying me and six other passengers and at times seemed utterly overmatched by the choppy waters of this stretch of ocean. Travelers disembark at the dock of Floreana's only population center, Puerto Velasco Ibarra, a small grid of volcanic gravel roads where they will find the island's few hotels and restaurants as well as its tourism center. Most visitors then board trucks operated by travel agencies for group tours, which are small semis with hollowed-out side-less trailers fitted with wooden benches. After reaching the island myself, I met Jorge, a Floreana local who works for Island Conservation, the NGO that has partnered with the Galápagos National Park to plan the ongoing effort to eradicate rats and feral cats. We climbed aboard Jorge's motorbike and followed the tour trucks up the road to the highlands of the island.

The ride was a short one, but the journey nonetheless amounted to a dramatic transition. As we climbed the steep dirt road, I felt the air get thicker against my skin and the temperature rapidly dropped several degrees. The vegetation shifted accordingly, from sparse shrubbery reminiscent of the hiking trails in the semi-arid climate of my hometown, Santa Barbara, California, to a dense green landscape befitting the heavy, humid air. We spent the rest of the day visiting a number of different construction sites, some completed, some in progress, where structures would house livestock or native animals during the duration of the planned poisoning program. We never came face-to-face with a rat during this tour, yet the imprint of this arena in humanity's global struggle against the rodents was evident.

In many ways, some glaring and some much more subtle, the frigid, snow-covered Albertan prairie and this verdant equatorial climate where we will spend these next two chapters provide a contrast that makes for ideal (literal and metaphorical) bookends to this exploration of rat control. The obvious contrast in the ecological, aesthetic, and cultural contexts of these two places makes for disparate conditions in which their shared project of extermination takes place. Of course, one key difference is that of underlying motivating factors (economic calculus versus environmental conservation). Another key difference, though, is the respective statuses of rats, both in the ecologies of either context and in their cultural imaginations. Whereas rats are famously absent in Alberta, yet loom large in the collective imagination there, in the Galápagos they could be said to occupy the inverse of these statuses. Rats are very much present, in a real, material sense, acting as troublesome agents in the islands' ecosystems for the environmentalists who hope to eradicate them. Yet, as cultural objects, they are far less individually salient here, buried beneath a sensory overload of other animal iconography devoted to iguanas, seals, myriad native bird species, and giant tortoises, as well as physically hidden in burrows beneath the lush landscape of Floreana's highlands.

Accordingly, it may have been rats that brought me to the Galápagos Islands to begin with, just as they brought me to Alberta and Los Angeles, but understanding rat control in this context requires special attention to the cultural meanings of other animals as well. In this chapter, I emphasize this fact by taking a relational approach to the animals that make up the diverse ecology of the islands, one that understands them, as cultural objects, in terms of each other. The species eradication programs in the Galápagos that attempt to conserve the habitats of native animals bring these important relationships to the foreground. More than that, they also reveal the extent to which we turn to animal species to make more coherent sense of the morally inflected, overarching set of values that structure our relationships to the environment.

This is because, in the Galápagos, the story of rat extermination is a story about nature, through and through. It is nature, more than anything else, that brings hundreds of thousands of tourists to the islands each year through the gates of Seymour Airport on the tiny island of Baltra, and nature that sends them fanning out from there throughout the archipelago, on day trips and guided tours, to bear witness to a variety of ecological spectacles. And it is this same nature that environmentalists are hoping to protect by systematically exterminating rats. Despite the ritualistic, deeply affective, and even spiritual character of the pilgrimage that so many make to see the nature of Galápagos, conversations about nature, and especially about protecting it, tend to rely heavily on appeals to scientific rationality and to define the ecological greater good in these seemingly detached, unemotional terms. At the heart of any discussion about conservation, though, lies a value system based fundamentally on shared cultural meaning. In this chapter, the project of rat eradication, and invasive species eradication more generally, serves as a window into these value systems, and through examining it I attempt to bring more clarity to the role of cultural meaning in environmental conservation. As we will see, the story of rat control being a story about nature means that the story of rat control is a story about tortoises, snakes, mockingbirds, chickens, cows, and, of course, human beings, and what kind of comprehensive whole they all make up as component parts.

In the rest of this chapter, I will examine species eradication programs as they relate to the cultural meaning surrounding the Galápagos Islands, the species at the center of the projects, and nature and the environment more broadly. I argue that species eradication programs intervene in two overlapping and interconnected ecologies: the physical, material ecology and what I refer to as the "ecology of meaning": the web of meaning and significance that structures how we relate to the nonhuman life in our proximity. This concept allows us to see both how elements of ecologies, like individual species of animals, become salient cultural objects and how those piecemeal significations combine to generate broader ideas of "nature," in much the same way as the material ecology is composed of a web of relationships and interactions between individual organisms. The next sections delve into the history of human activity in the archipelago and how its animals and natural landscape came to hold such iconic symbolic meaning. This background, along with the broader cultural history of environmentalism, informs the case study of species eradication on the Galápagos Islands, including the current rat eradication on Floreana as well as past eradication programs of different species.

These programs on the Galápagos reveal insights into the relationship between individual animals as meaningful cultural objects and more holistic cultural ideas of nature. These overarching ideas of nature are potent motivations for species eradication work, but these ideas are most clearly articulated in reference to the symbolic meaning of individual animals. Moreover, these ideas are negotiated through discourses surrounding the question of acceptable environmental change: Who or what can be a legitimate agent of change in an ecosystem? This discourse is especially salient for conservation work in the Galápagos because the islands are iconic for their association with Darwin's theory of evolutionary change, and also because the scale of these conservation programs makes them significant human interventions in an ecosystem's balance of life. This question of acceptable change serves as a net in which we might capture more nebulous yet potent cultural discourses, offering a better understanding of the nuanced relationship between the materiality of ecosystems and their place in the cultural imagination.

HUMAN HISTORY IN THE GALÁPAGOS

Invasive species eradication for conservation is used in many parts of the world, but especially on islands, where relative isolation makes it more feasible. The Galápagos Islands are an ideal site to examine this conservation method from a sociological perspective, given how iconic they are for their nature and nonhuman life. Also, the islands were home to one of the most well-publicized and controversial species eradication programs to date between 1997 and 2006. Referred to as "Project Isabela," this was an eradication of goats to protect tortoise habitats. The organization Island Conservation is now aiming to eradicate rats and feral cats on Floreana Island in order to save native bird species as well as to potentially reintroduce other animals that once lived there but no longer do. The organization is more broadly dedicated to conserving native species on islands, primarily through the eradication of invasive species, and has programs in locations across the Pacific and Atlantic Oceans as well as the Caribbean Sea. They have partnered multiple times with the Galápagos National Park and the Galápagos Conservancy to assist in performing eradication programs on various islands in the archipelago. When I visited in December 2018, Project Floreana was several years in, though "implementation," or the beginning of the baiting and poisoning phase of the program, was then still many years away. The project faces unique challenges that help foreground tensions between nature and culture; it is the first rodent eradication program of this scale on an island with a significant permanent human population.

When I would tell people about my plans to visit the Galápagos for this project, I was surprised by a particular trend in these conversations; while I never encountered anyone who had not heard of the islands, many people told me they had not realized there was a human population there. For me, this confirmed both the status of the archipelago as an iconic symbol of nature and the endurance of definitions of nature as untouched by human hands. It is also a good reason to explore the human history of the Galápagos that provides the backdrop for contemporary conservation efforts there. As much as the material, biophysical elements of the islands, this history of human activity in the Galápagos is responsible for turning them into the bastion of the natural world that they now are. In fairness to those people I spoke with who assumed the only human presence on Galápagos came in the form of scientific researchers retracing Darwin's footsteps or intrepid, nature-loving tourists seizing the chance to photograph the islands' famous tortoises, the human population on the archipelago dates back only a few hundred years. Moreover, the islands were home to fewer than 2,000 residents at the time of Ecuador's first national census in 1950. That figure has grown rapidly in the decades since, and the 2010 census counted a total population of 25,124.[1]

The first documented human visitors on the islands arrived there by accident in the sixteenth century, having been blown off course by the South Equatorial Current (Nicholls 2014). For the next several centuries, the Galápagos served as a waystation for whalers and buccaneers. Like so many of those who travel to the islands today, these early visitors were drawn to the islands by the giant tortoises for which the archipelago is named. However, in this early period of the human history of the Galápagos, tortoises were a material commodity rather than a natural marvel to behold. For seafarers stopping between long stretches on the open ocean, the tortoises were a valuable source of food. Tortoises were harvested both for their meat, which was grilled or cooked into soup, as well as their fat, which served as an alternative to butter. This was the case when the most famous visitor to the Galápagos arrived in 1835. Like the other voyagers who stopped on the Galápagos in this period, Charles Darwin and the rest of the crew aboard the HMS *Beagle* provisioned themselves with tortoises to eat during and after their stay in the archipelago (Nicholls 2014).

To some encountering the details of Charles Darwin's famed visit to the Galápagos for the first time, the fact that the islands' iconic tortoises served as an important source of food for the famed naturalist may seem perverse, perhaps even to the point of being sacrilegious. Ultimately, understanding the current state of the Galápagos—their allure to travelers and the drive of conservationists to protect and restore them at great effort—means comprehending how their giant tortoises went from foodstuff to cultural icon. This transition in symbolic meaning unfolded in the context of both the human-nature interactions that have occurred on the Galápagos Islands in the years

---

1   Figures taken from the website for Ecuador's National Institute of Statistics and Census, https://www.ecuadorencifras.gob.ec/estadisticas/.

since 1835 and broader cultural forces beyond that saw the development of environmentalism as we understand it today.

One aspect of human activity on the islands that contributed to this symbolic shift for tortoises and other endemic species in the Galápagos, of course, is the influence of third-party organisms: introduced species that have arrived there with humans. Various species of rats have found their way to the islands aboard ships at different historical periods. Black rats (*Rattus rattus*) are believed to date as far back as the sixteenth or seventeenth century in the Galápagos, having arrived on whaling ships. The brown rat (*Rattus norvegicus*) appeared as the population of the islands boomed in the 1980s. These rodents are responsible for much of the habitat destruction and other deleterious ecological effects that have put various endemic species at risk of going extinct and have prompted ambitious conservation programs. Other introduced species that have imperiled tortoises, native birds, and other animals are livestock. The effects of agriculture and cattle grazing on the landscape are often cited as a stressor on the ecology that threatens the islands' pre-human wildlife. Feral goats that spread beyond the human-populated parts of the Galápagos are also a concern for conservationists.

These are the material conditions that account for the current situation in the Galápagos. Humans, with help from rats, goats, cattle, feral cats, and other species they have intentionally or unintentionally introduced to the islands, have dramatically altered ecological conditions there in a relatively short period of time. The tension between people and the nonhuman landscape has only increased as the human population of the Galápagos Islands has dramatically grown since the 1950s and as the tourism industry has likewise exploded. While agriculture, fishing, and other industries that extract material resources from the islands' ecosystem continue to be in tension with conservation goals, there is also a widespread consensus on the importance of preserving the Galápagos's unique ecosystem and saving its species from extinction. While these material circumstances help explain why various species are imperiled, though, they do not provide a full picture of how those species became iconic symbols of nature and how their preservation became such a priority. How did we get from Charles Darwin eating grilled tortoise meat and tortoise soup to the Charles Darwin Research Center, along with the Galápagos National Park and other organizational partners, spearheading efforts to save those same tortoises?

Charles Darwin and the Cultural Myth of the Galápagos

The legacy of Darwin himself on the islands is of paramount importance. Iconography devoted to him is ubiquitous during a trip to Galápagos, from the businesses and souvenirs that bear his name or likeness available in Puerto Ayora, the islands' largest population center, to the town's main drag, named Avenida Charles Darwin. Darwin serves as an avatar for nature at large, his theory of natural selection having cracked its code and rendered visible its mysterious logics and mechanisms. These associations have transformed the Galápagos into a secular temple and elevated the animals that helped inspire Darwin's thinking to the status of revered idols. This is crucial context for environmental conservation efforts there. While justifications for eradicating rats, goats, or feral cats might lean on scientifically rational notions like biodiversity and a greater ecological good, the endemic tortoises, finches, iguanas, and other species of Galápagos that stand to benefit from these programs carry immense symbolic value that also generates popular support for them.

For an informal but illuminating illustration of the cultural mythology surrounding Galápagos and Darwin, the reader can use one of the many AI image generators that have become increasingly widespread in recent years and enter the prompt "Charles Darwin in the Galápagos Islands." Chances are, the visual representation this experiment will yield, which is generated from the prompt based on training data gleaned from the internet, will depict an immediately recognizable Darwin foregrounding a landscape of beaches, outcroppings of volcanic rock, and greenery. While these images will likely conform to most people's first associations with the prompt fed to the AI engine, it is also likely that they contain a specific historical inaccuracy; the Darwin depicted in these renderings typically sports a prodigious gray beard and a brow furrowed from years of deep meditations on the complexities of nature.

In reality, the Darwin that visited the Galápagos Islands was a young man, only twenty-four years old when the HMS *Beagle* arrived there. Nearly two centuries later, though, Darwin's older self is the one that is most indelible in the popular imagination, as crudely approximated by the algorithmic processing these AI models do. Given the fact that his most iconic and influential work, *On the Origin of Species*, was not published until over two decades after the *Beagle*'s circumnavigation of the globe, it is not surprising that the version of Darwin's visage that is best remembered is him later in life. The effects of time have compressed the story of Darwin, the Galápagos, and what he learned there, eliding the years that elapsed between his observations there and his synthesis of them into a theory of evolution. It is not only AI-generated images that reflect this compression. The cover for the entry devoted to Darwin in the bestselling "Who was . . . ?" series of autobiographies published for children by Penguin is not unlike the average AI rendering based on the prompt above. That cover, which likely has served as the very first introduction to Darwin for many US children, shows the wizened, bearded version of Darwin jotting down notes and illustrations of the giant tortoises that surround him, with the HMS *Beagle* in the background.

In the grand scheme of things, this trend amounts to a fairly minor historical inaccuracy, but it sheds light on the broader relationship between cultural meaning and the Galápagos. Darwin himself is synonymous with evolution, and, though his magnum opus was published half a world away and decades after his visit there, the Galápagos Islands have become synonymous with him. The power of this cultural mythology, in which the tortoises and finches that helped reveal the laws of natural selection have accordingly become symbols of evolution itself, helps account for some of the cognitive dissonance the islands might engender. Because the islands are hallowed ground for nature and its extra-human logics, we might forgive those who knew little about the islands aside from those associations and therefore were surprised to learn that the archipelago has a substantial, permanent human population. Because Galápagos tortoises have such cultural significance, it is understandable that some would not have guessed that Charles Darwin once subsisted on their meat. But the development of this cultural myth, and thus these sources of cognitive dissonance, relied on much more than just Darwin and the islands themselves. It is also grounded in wider notions of nature that sprang from the development of Western environmentalism between the voyage of the *Beagle* and now.

The "Hows" and "Whys" of Saving Nature

A wide variety of local, national, and international organizations are involved in managing the Galápagos Islands and stewarding their environmental protection. Notable among these are the Galápagos National Park, established in 1959 by the Ecuadorian government, and the United Nations Educational, Scientific, and Cultural Organization (UNESCO), which has designated the islands as a World Heritage Site and established a Marine Biosphere Reserve that covers an area of roughly 133,000 square kilometers, encompassing the waters that surround the archipelago. These different conservation organizations, along with a plethora of nongovernmental organizations that collaborate on various initiatives, reflect a diversity of approaches to environmental conservation and underlying philosophies regarding nature and what it means to protect it.

The Galápagos National Park, Ecuador's first such designation, has roots in the early American environmental movement that spawned the proliferation of national parks first in the United States and later across the world. The impetus for the first national parks was concern that industry and human activity threatened to spoil the majesty of places deemed national treasures. This sentiment grew in part out of the cultural movement of Romanticism, which placed an emphasis on the aesthetic wonder of the natural world and the powerful emotional responses elicited by the "sublime." This early incarnation of the conservation movement was spurred by deep, earnest desires to commune with the natural world and to protect humanity's capacity to do so. Given the importance of Romanticism's aesthetic sensibilities, though, these desires were applied unevenly to different landscapes and locations. As William Cronon notes in his classic essay "The Trouble with Wilderness," it took until the 1940s for the United States to designate a national park in a swamp ecosystem (Everglades National Park), and it still has not established one in the grasslands (Cronon 1996).

This uneven distribution of protected areas points to one of the most prominent criticisms of so-called fortress conservation, which is epitomized by national parks. This mode of conservation attempts to parcel off areas that are spared from development and other transformative effects of human industry. While this is still the core notion underlying many of the conservation efforts that exist today, environmentalists have increasingly found this approach to be inadequate on both practical and philosophical levels. For one, it has become increasingly clear that sequestering nature from damaging forms of human influence is much harder than simply drawing geographical boundaries around areas to be protected. Environmental icon John Muir, often deemed the "Father of the National Parks," famously wrote that "when we try to pick out anything by itself, we find it hitched to everything else in the universe." By this very principle of interconnectedness, it is evident that the confines of a national park cannot possibly be completely cordoned off from the industrial pollution of a factory outside park boundaries or, for that matter, the greenhouse gas emissions of a global economy powered by fossil fuels.

Moreover, fortress conservation rests on an ambivalent conception of humanity's own role in and relationship to nature. On one hand, national parks are conserved for the enjoyment of human visitors, which accounts for their uneven distribution across different ecosystem types that Cronon attributes to cultural and aesthetic preferences. On the other hand, though, the idea that such places need to be protected from the destructive forces of human influence implies that humans themselves have no role to play in nature, at least no positive one. These criticisms have inspired alternative approaches to the conservation of nature that attempt to embrace interconnectedness and bring human activity into harmony with the natural world rather than artificially separating the two. This emphasis on "sustainable development" is the underlying principle of UNESCO's Biosphere Reserve program. In the Galápagos, the steadily increasing human population coupled with the ever-fervent popular support for conservation efforts there has necessitated this approach.

The contradictions stemming from the interests of various human stakeholders in the Galápagos came to a head in the 1990s. The primary driver of immigration to the islands has always been their robust tourism economy fueled by the desire of travelers from around the world to experience their ecological wonders. This population boom and the ancillary industries, especially fishing, that tourism has spawned pose threats to the islands' native species and their habitats. Fisherman and other Galápagos businesspeople clashed with conservation organizations in a series of incidents between the mid-1990s and the early 2000s that occasionally involved vandalism of facilities operated by the National Park and the Charles Darwin Research Center. In some instances, protesting fisherman used giant tortoises as leverage, threatening to harm them if their demands for greater fishing rights were not met (Brown 2004). One outcome of this conflict was a landmark piece of legislation called the "Special Law" enacted in 1998. This law attempted to chart a path toward sustainable development in the Galápagos that satisfied both fishermen and conservationists. It expanded the Marine Reserve, which increased restrictions on commercial fishing, while also affording residents of the islands more political autonomy (Hoyman and McCall 2013).

The debates surrounding how best to steward nature are far from settled. The conflict between conservationists and fishermen indicates how messy the problem of conservation can be, even in an archipelago where 97 percent of the land is a national park. In a *New York Times* article from 1995 reporting on these tensions, a scientist at the Charles Darwin Research Center gets at the heart of this difficulty. Referring to the influx of immigrants to Galápagos from the Ecuadorian mainland and elsewhere who seek employment or entrepreneurial opportunities in the islands' tourism industry, he is quoted as saying that "the new people who are coming don't even realize what the Galápagos means to the rest of the world" (Schemo 1995). While many would have likely countered this by pointing out that the importance of the Galápagos to world travelers is precisely why such newcomers arrived on the islands, this quotation nonetheless rightly frames the conflict as one over "meaning." The rights of Galápagueños to consume natural resources, the intrinsic value of endemic species, and the most ethical and appropriate way for humans to enjoy nonhuman nature are all questions that lie beneath conflicting sets of cultural beliefs about the value of nature that play out through conservation efforts. Programs like the rat eradication effort on Floreana Island have concrete goals, but implementing those goals ultimately involves wrestling with the questions of nature's value and the proper role of human influence in an ecosystem like the Galápagos. Given this context, eradication programs and other conservation efforts not only make material interventions in the ecosystem, they are expressions of values, entries into the broader discourse about the meaning of nature.

THE ECOLOGY OF MEANING

To understand the struggle to conserve nature in the Galápagos, it is crucial to take the cultural mythology of the islands, their history, and their nonhuman species seriously. A simplistic reading of the conflicts between industry and conservation there might cast it as a story of "jobs vs. the environment." This narrative frame, where the material livelihood of one group of stakeholders is at odds with the scientifically rational imperative to preserve biodiversity, recurs frequently across various arenas of environmental conflict, such as the transition from fossil fuels to renewable energy sources. But in the Galápagos, the iconic symbolic value of nature shifts the gravity of these debates among different stakeholders. The industries that pose threats to the ecosystem are nonetheless dependent on the tourism economy centered on the island's natural wonders. The tourists who romp around the islands on day trips and backpacking adventures likely do not return home and regale their friends and family with tales of biodiversity and ecosystem services. While the value of nature may be linked to concepts like these and the popular recognition of their importance, the Galápagos Islands are world famous for their symbolic meaning, not for any abstract ecological concept they epitomize.

As such, I analyze species eradication programs in the Galápagos as negotiations of this symbolic meaning. By intentionally intervening in the ecosystem, these programs not only affect the material balance of the islands, they simultaneously manage another interconnected system, what I refer to as the *ecology of meaning*. This latter ecology encompasses the systems of shared meaning that shape our attitudes about and valuation of the nonhuman life in our proximity. In other words, the ecology of meaning represents our capacity to construct "the rat," for example, in our collective cultural imaginations as a figure that exists somewhat independently from the materiality of rat bodies and lives themselves. This concept calls attention not just to species of animals as cultural objects, but to the relationship between those cultural objects and how they add up to a comprehensive cultural understanding of the nonhuman environment.

By referring to the ecology of meaning as a symbolic structure overlaying an ecology in the traditional sense, I approach conservation and species eradication as a problem of the evolving relationship between materiality and meaning. This builds upon scholarship in cultural sociology that has examined the conditions under which material things are integrated into the symbolic arena of social life, or how and why they do or do not become "resonant" as cultural objects (McDonnell et al. 2017). For instance, McDonnell investigates HIV/AIDS media campaigns in Ghana and the frequent and troublesome dissonance between those campaigns' intended meanings and how they are received by their audiences (2010, 2014, 2016). Condoms become repurposed as fashion accessories, the intended messages of dilapidated billboards are covered up by fliers, and the red dye of HIV/AIDS awareness ribbons fades in sunlight, transforming them into pink breast cancer ones. With these examples, McDonnell demonstrates the dynamic and delicate relationship between materiality and cultural meaning, a relationship captured by his term "cultural entropy," which he defines as "the process through which the intended meanings and uses of a cultural object fracture into alternative meanings, new practices, failed interactions and blatant disregard" (McDonnell 2016: 26). The fraught relationships between materiality and meaning described here are especially applicable to plants and animals, whose existence independent of the human regimes of meaning-making in which they participate make them especially "unruly" (Domínguez Rubio 2014). The "environment" essentially comprises a special category of material object in the popular cultural imagination, defined by an assumed autonomy of sorts from the human social world. This ontological separation notwithstanding, elements of the natural world function as cultural objects in essentially the same way that material culture like pieces of artwork or consumer products do, as described by scholars of material culture.

My approach builds especially on the work of Fernando Domínguez Rubio, answering his call for an "ecological" approach to the study of material culture. By ecological, Domínguez Rubio means one that focuses on the "processes and conditions" under which material things may become vectors for meaning (2016). He further highlights the ephemerality of these relationships, noting that the materiality of things is in constant flux, and so too is their capacity to hold specific meanings. Domínguez Rubio makes an analytical distinction between "things," which he defines as "material processes that unfold over time," and "objects," which are "the positions to which those things are subsumed in order to participate in different regimes of value and meaning" (2016: 61–62). To think about material culture and meaning "ecologically," then, is to analyze the relationship between things and objects, or the conditions under which things can function as objects.

I apply such an approach in this chapter to specifically environmental topics and expand what is meant by "ecology" accordingly. In one sense, the ecology in question is metaphorical in the ways suggested by Domínguez Rubio, but it simultaneously refers to ecology in the traditional, literal sense. In other words, whereas a material ecology is made up of its component parts and the totality of the relationships between them, I argue that the cultural meaning of that ecology is also composed of meaning on the smaller scale of individual cultural objects. Domínguez Rubio's approach is useful for examining the capacity of certain species to function as cultural objects, but in applying and expanding this framework, I show how that meaning-making generates more holistic, overarching understandings of nature and the environment that in turn frame struggles over and discourses surrounding conservation efforts like those in the Galápagos.

Our relationships to nonhuman environments are best understood in light of both the delicate balances of material ecologies and the equally delicate balances of ecologies of meaning. Moreover, when it comes to our interactions with the physical environment, the conditions of these two ecologies are inextricably linked. In other words, when human intervention impacts a biophysical landscape, it necessarily affects the material balance of life there and the meaning those life-forms carry or are capable of carrying, their "resonance" (Kubal 1998; McDonnell 2014; McDonnell et al. 2017; Robnett 2004) as cultural objects. Thus, when a conservation program acts to protect a certain species and likewise to exterminate another, it makes simultaneous and parallel interventions into the material ecology and the ecology of meaning. Understanding conservation programs in these terms allows us to see how the project of environmentalism itself is implicated in much broader regimes of ethics and morality that extend well beyond typical notions of what constitutes an "environmental" issue. The stakes of species eradication programs are thus not simply which species should be given special protections and which ones should be erased from a landscape, but the related notions of who and what is deemed venerable or expendable.

A central cultural discourse of concern to conservationists is the question of acceptable versus unacceptable environmental change. The negotiation of this question functions as a distillation of a broader and messier cultural value set associated with "nature." In other words, the discussion of change, which became a particularly salient thread throughout my interviews and fieldwork, serves as what Ghaziani calls an "observable analytical unit" (2014) through which the broader cultural meaning under examination becomes clear. Probing this area of negotiated meaning will provide an index into the connections between animals as cultural objects and overarching notions of nature that make up the ecology of meaning.

STATIC SYMBOLS OF CHANGE: THE RELATIVE AUTONOMY OF ANIMALS' CULTURAL MEANING

While the Galápagos Islands are themselves emblematic of unfettered nature and a brilliant diversity of life, some particular animals serve as the most recognizable cultural avatars of that nature. These include birds like the famous Galápagos finches, who boldly perched next to my plate on the lunch table during my stay on Floreana Island, the frigate birds, whose males are instantly recognizable by their inflatable red gular pouches, and the eccentric blue-footed boobies, whose otherworldly appearance and silly-sounding name inspire a robust market of PG-13-rated souvenirs on offer in Puerto Ayora.

Other native and endemic species like marine iguanas, lava lizards, and fur seals all captivate audiences of tourists as well. These animals are the all-star cast of characters in the ecological theater that the Galápagos Islands are stage to. The tourists I encountered while visiting the islands tended to talk about their visits specifically in terms of the animals they hoped to see. One traveler who boarded the tiny boat to Floreana Island with me told me he was making the two-hour journey specifically "because he wanted to see a penguin." On a day trip to North Seymour Island, a "birder" couple listed the avian species they hoped to add to their lists there, some of which they photographed so eagerly as to draw repeated admonishments from our guide to keep a greater distance.

It is unsurprising, then, that individual species often serve as the focal point around which many conservation efforts, including eradication programs, are organized. Many of the most visible such efforts, like Project Isabela's goat eradication, center on the islands' most iconic animal, the Galápagos giant tortoise. The programs aimed at saving these animal characters from extinction and the broader fascination with (and affection for) them prompt us to consider the relationship between the animals' material existence as biophysical entities in broader ecologies and their status as cultural objects.

The symbolic cultural meaning that these animals carry is deeply bound up in the moral discourses of environmentalism. Protecting particular animal species is the most tangible way to operationalize the affective and moral impulses of an environmentalist ethos which, when taken more holistically, become much messier and fraught with contradictions. Often, during my interviews with species eradication practitioners, interviewees affirmed the overarching importance of saving endemic species, even if they ultimately acknowledged that the basis for this was a value judgment as much as a rational truth of ecological science. For instance, Chris, an organizer of Project Floreana for Island Conservation, discussed the position some opponents of species eradication take, that killing introduced species to save others is a recklessly arbitrary decision and even one that amounts to "playing god" in the ecosystem. When this topic came up, Chris surprised me by conceding, "Yeah, they've got a valid point." Nonetheless, he remained firm that "when it comes down to it, we're not embarrassed to say we value an endemic species more than an introduced species that isn't at risk of going extinct." Though some vocal opposition exists, especially from animal rights groups that object to the brutality of species eradication methods and even more generally to the use of killing as a tool of conservation, the value Island Conservation places on endemic and native species is widely shared and their efforts have enjoyed considerable support.

Brian, a pioneer in the field of species eradication whom I spoke to prior to visiting the islands, humorously demonstrated the extent of this cultural value favoring native animals in a different way. "Make sure you pack a flashlight," he warned me, before explaining that, should there be a power outage where I was staying, it would be unwise to get out of bed in the dark because of "the centipedes." After he detailed the unpleasant effects of these creatures' poisonous bite (not fatal, but "after the second day of fever you'll *wish* you were dead"), I joked that maybe Island Conservation could eradicate them as well. He dismissively said, "Well, they're native," though he then further emphasized the terror these centipedes inspire: "Oh, they're so scary. Remember all the shit that was coming out of the canyon walls in *King Kong*? It's like that." Surely these centipedes are not what comes to mind for most when they picture the iconic wildlife of the Galápagos, but Brian's distaste for them comes across as more of a respectful terror compared with the disdain he expressed in other parts of our conversation for the rats and goats that are the targets of eradication.

The moral basis for conservation programs puts the terrifying Galápagos centipede in the same category as the Galápagos tortoise (as native wildlife to be preserved), but the tortoise is of course a much more charismatic figurehead for these efforts. Just as they are protagonists of tourists' encounters with nature, they are also the sympathetic protagonists of projects like the goat eradication on Isabela Island. In this way, these charismatic animals function as important cultural objects in ways that transcend their material existence. Of course, it is impossible to completely decouple animals' cultural meaning from their materiality, and this is uniquely true in the Galápagos Islands, where, thanks to the cultural mythology of

Darwin's visit there, these visible animals are symbols of the broader ecological processes within which they exist. Nonetheless, animals' cultural meaning remains relatively autonomous from their materiality in ways that expose contradictions in the relationship between the two.

Arguably, the concept of change, more than anything else, is what makes the Galápagos Islands and their animals as iconic as they are and is at the center of the moral ethos that privileges native animals over introduced ones. Tourists flock there to see an abundant and thriving ecosystem up close, but that ecosystem's special allure stems from its enduring association with Charles Darwin's research there and his theory of evolution that it inspired. As one of the animals most closely associated with Darwin's legacy (it even provides the logo for the Charles Darwin Foundation), the Galápagos tortoise is potently symbolic of the adaptive change that is possible through evolution and natural selection. This association comes from observations of the physical differences that have developed in tortoises on different islands according to the specific conditions of the landscape. As a result, the tortoise has become a cultural metonym for the very idea of evolutionary change and its cultural significance.

While the tortoises themselves may be involved in a long-standing and ongoing process of evolutionary change, their status as symbolic objects representing that change is static by comparison. Species eradication programs throw this tension into relief. By tinkering under the hood of an ecosystem, they strive to maintain the material conditions under which the tortoise retains its resonance as a cultural symbol of evolutionary change, though at some level this inherently alters the unfettered quality of natural selection that is central to those conditions under popular ideas of nature.

This points to the importance of cultural meaning as a driving force behind species eradication, and to that meaning's messy relationship with materiality. Species eradication advocates and practitioners often express their motivations for conserving species like the Galápagos tortoise in terms of preserving the material process of evolution itself. As Brian expressed it, if one of these species were to die out, it would represent a loss on a higher order: "You're not only losing species . . . you're stopping evolution in its tracks in that system. Because once you lose a species, that unique genetic combination is gone." Species eradication programs are not just attempting to save species, they are attempting to protect the integrity of evolution itself, which should not be compromised by forces like human economic activity or the spread of introduced species that comes with it. The practical question underlying all of this is what amounts to acceptable, as opposed to unacceptable, environmental change, particularly when humans are involved. According to the cultural negotiation that conservationists must undertake, humans can be legitimate agents of change insofar as they are acting to preserve a specific imagination of the material landscape. In a sense, more than preserving evolution itself, the driving cultural impulse behind conservation programs is to preserve the iconography of evolution, as exemplified best by tortoises and finches. Put a different way, while the theory of evolution is fundamentally about dynamism and change, the cultural mythology of evolution has invested significant meaning in a temporal snapshot of Galápagos and its species as they were when they inspired the theory and were thus immortalized as cultural objects. Of course, this is not to say that the efforts are disconnected from these ecological principles, but it is important to acknowledge that they are motivated not just by the notion of evolution itself (or by other ecological principles like biodiversity), but also by the cultural importance of the concept, as represented by the iconography associated with it, in other words, the static symbols of change.

It is the potency of the cultural meaning of these symbols that makes the broader moral distinction of native versus introduced species salient. Whereas native species symbolize acceptable change in the form of evolution, the rats of Galápagos, having stowed away on the vessels of human interlopers, are themselves emblematic of the wrong kind of change. Just as their symbolic meaning hovers above their material existence, the meaning of "nature" as a cultural object in its own right is similarly autonomous from the materiality of ecologies where we might locate it in place. The next section turns to the relationship between these two orders of symbolic meaning, examining how piecemeal significations like animals as cultural objects are the building blocks of more holistic ideas of nature.

ANIMALS AND NATURE: PIECEMEAL AND HOLISTIC CULTURAL MEANINGS

As explored above, species eradication programs in the Galápagos negotiate the cultural boundaries of acceptable environmental change. This includes the central question of humans' roles in ecosystems and the extent to and conditions under which they can be legitimate agents of change according to prevailing cultural values. This negotiation happens in the dynamic relationship between the cultural meaning of animal species and overarching cultural ideas of "nature." For example, the notion of evolution symbolized by native species is one of undisturbed evolution, which rests on notions of a fundamental separation between human society and nonhuman nature. This particular ideological tendency, as discussed in chapter 4, is often referred to as the "nature/society divide" and permeates seemingly all discussions of environmental matters. It has been examined and critiqued by environmental sociologists in the past (Catton and Dunlap 1978, 1980; Dunlap and Catton 1983; Freudenburg et al. 1995; Murphy and Dunlap 2012). In the Galápagos, these sensibilities around nature dictate that humans should not have a hand in influencing the course of evolutionary change, but humans have done just that, in large part by intentionally or unintentionally introducing invasive species. As cultural objects, then, the invasive animals of Galápagos, like the rats of Floreana Island which are thought to have originally arrived as stowaways on the ships of pirates and whalers, represent disruption of acceptable change as proxies for unnatural human influence.

In Project Isabela's goat eradication, the nature/society divide was symbolically and spatially represented by a particular feature of the island's landscape. The island is formed by six shield volcanoes, two of which form the oblong southern portion of the island, and the rest of which extend in a narrow strip to the northwest. These two sections of the island are separated by an isthmus of volcanic rock. This area, named the Perry Isthmus, forms a natural barrier between the human-inhabited south and the northern part of the island where the eradication took place. For this reason, the isthmus was key to preventing the human activity in the south (and the goats that serve as proxies for that activity) from compromising the long-term goal of returning the north to an undisturbed "pristine" condition (Charles Darwin Foundation and Galápagos National Park Service 1997). Moreover, this barrier can be seen not just as separating the realm of nature from the realm of society, but also as delineating legitimate human-generated ecological change from unacceptable such change; the effects of the human population on the south must be mitigated by the human intervention of species eradication in the north.

One aspect of the more recent eradication program on Floreana Island illustrates this distinction particularly well. Carmen, an environmental lawyer working on the program, described to me the way the conservationist team's relationship with the population of Floreana has had to evolve over the course of the program with this story about its early days:

> CARMEN: The ideal for a project like this is not having people and not having [farm] animals at the moment of the eradication. And the eradication really lasts three months. So, the first idea was "OK we're taking the hundred and fifty people out and we're taking all the cattle and everything out of the island, we intervene on the island, and then we bring you back."
> AUTHOR: Really? Wow.
> CARMEN: And they said, "No way."
> AUTHOR: "Absolutely not."
> CARMEN: "Absolutely not!" Well, [that way] you don't need to mitigate the risks for people, you don't need to make any plans for mitigation of animals . . . and the investment of the project is lower.
> AUTHOR: Wait, to be clear, that was actually proposed in the beginning?
> CARMEN: Yeah.
> AUTHOR: Oh, wow. . . .
> CARMEN: That's . . . logic for us. It's like, that's scientific.

Eventually those in charge of the project scrapped the plan to remove the human population from the island during the eradication and has worked much more collaboratively with them since. But, in light of the historical tensions between conservationists and the permanent population of the islands, it is nonetheless telling that this was the initial "scientific" assumption—that an eradication program would necessarily entail removing human residents for the duration of the program. It speaks to a fundamental incompatibility that is assumed between activity by (non-conservationist) humans and the goals for nonhuman nature there. As in Isabela, though, the eradication program on Floreana involves a massive human intervention in the landscape. For instance, I visited the sites of aviaries under construction for the awe-inspiring undertaking of housing the population of a native bird species for over a year in the run-up to and during the eradication program. Other structures intended to sequester livestock during the operation were still in progress. Moreover, like the Perry Isthmus in Isabela, procedures like these clarify the boundaries of legitimate environmental change according to the nature/society divide and express those boundaries in space.

All of this is to say that there is a discernible, overarching moral logic of environmentalism that guides the eradication effort, tied to a specific imagination of the "nature/society divide" and visible through the negotiations of acceptable environmental change. Importantly, though, these holistic ideas of nature that lie at the heart of eradication programs are difficult to distill into a single coherent ethos. The contours of those ideas are thus clarified through references to and relationships with symbolically meaningful individual animals. In my interviews with people involved in the programs, I attempted to construct a working understanding of these overarching cultural values vis-à-vis nature by asking them to describe their end goals for the programs—for instance, what was their vision of Floreana Island when the eradication program was over? Would it return the landscape to a state reminiscent of a particular point in history? Or create a different kind of human-nonhuman balance in the ecosystem? Perhaps the most revealing theme that came out of this line of inquiry was a general lack of specificity. For example, when I posed the above questions to Kevin, the director of that project, he responded: "So, all of those things are values based. . . . So part of this for my bit is understanding people's values. . . . And then looking to have those reflected . . . in the visions that you set for this stuff, because when you have those reflected out there people stick with that. . . . And it's more about that vision. And so this is what enables you to continue, as you're going through a project like this, [to] keep pointing to that vision. You know, 'You said Floreana would be a better place without rodents and cats.'"

Kevin declines to offer many specifics in terms of the material conditions of the island, but instead acknowledges that the end "vision" for the island once the eradication program is complete is grounded in cultural values rather than simply in scientifically rational notions of what a healthy ecosystem looks like. He does not directly elaborate that vision or those values, but he offers one concrete statement: that the project is predicated on the assumption that "Floreana would be a better place without rodents and cats." When "nature" is too big a cultural object to look at all at once, we understand it by turning to more discrete and visible cultural objects, the animals that we relate to directly. Carmen, meanwhile, imagined the future of Floreana this way: "Imagine being able to . . . [see] hawks flying around, or have snakes that were extinct, or 'Oh, see that turtle going around? That is because of the restoration project.'"

The Charles Darwin Foundation/Galápagos National Park Service reports compiled in advance of both Project Isabela and Project Pinzón (a rodent eradication on Pinzón Island that was completed in 2012) have the stated goal of returning the landscape to a "pristine" condition. The Project Isabela report goes into slightly more specificity: "The project goal, to restore the island to as near a pristine condition as possible, implies restoring fully the biodiversity of Isabela and its natural ecological and evolutionary processes, in the absence of introduced species" (Charles Darwin Foundation and Galápagos National Park Service 1997). The quotations above, though, indicate how operating definitions of "pristine" are understood in terms of animals' cultural meaning, from the rats, cats, and goats that symbolize defiled landscapes and interference with legitimate environmental change, to the iconic native animals that represent the cultural ideal of extra-human nature.

Species eradication practitioners are not simply intervening to save the material, corporeal existence of native animal species, they are managing their capacity to serve as symbols of a particular imagination of nature, much in the same way as the Louvre preserves the *Mona Lisa*'s ability to hold its own symbolic capacity, as Domínguez Rubio describes. The material practice of eradication work then is motivated by an environmental ethos of which it is also involved in the ongoing production. Animals themselves, though, serve as prisms through which we capture and distill the messier, sometimes incongruous aspects of this ethos. As we will explore in the next chapter, one of the central difficulties in maintaining and acting on a balanced moral worldview in conservation projects like these is the brutality of eradication work itself and how it contrasts with the affective motivations that lie at the core of this work.

THE MATTER AND MEANING OF NATURE WORTH SAVING

In this chapter, I have used the project of species eradication for conservation in the Galápagos Island as a case study to examine the place of cultural meaning in the multispecies ecologies that humans participate in. Through the lens of the ecology of meaning, I have attempted to demonstrate the dynamic relationships between the materiality of ecosystems and the cultural meaning they hold, both in terms of animals as cultural objects and more holistic notions of nature. The eradication of an invasive species for conservation purposes makes both material and cultural interventions in the landscape. By that, I mean eradication is intended to materially alter the ecological balance of the islands, but part of the function of this intervention is to preserve the conditions under which species of animals and the broader landscape they inhabit function as resonant cultural objects. The framework of the ecology of meaning helps us visualize the structure of this cultural meaning in its micro- and macro-forms as well as its relationship to the material ecological conditions of the islands.

This analysis charts connections between work that has analyzed animals as "totemic" cultural objects, work that has examined the material interconnectedness of human social life with multispecies ecologies, and work that has theorized the cultural significance of the concept of "nature" in the broadest sense. Generating a more robust picture of the connections between these promises to clarify our understanding of how environmental issues are implicated in social, cultural ones. For instance, in the Galápagos, the connections between overarching ideas of nature, the cultural significance of animals, and the material realities of ecologies play out in discourses of moral worth that adjudicate between what is killable and not, or who or what has a legitimate right to exist and where. These are cultural discourses with high stakes and far-reaching implications. We need not look further than the assumption that an island's human population must be removed from their homes for a conservation program to see how environmental issues are firmly entangled in the social implications of these discourses.

× seven ×

KILLING FOR LIFE

*Morally Acceptable Lives and Deaths
in Environmental Conservation*

For a period in the early 2000s, the sound of environmental conservation on the largest of Ecuador's Galápagos Islands was, give or take Richard Wagner's "Ride of the Valkyries," a lot like the audio from the most iconic scene of the film *Apocalypse Now*. The hum of whirring blades from helicopters circling above Isabela was punctuated by the staccato crack of gunfire from high-powered rifles. The targets of these aerial assaults were goats, and their death by firing squad was a sentence handed down by a group of conservation organizations including the Galápagos National Park Service, the Charles Darwin Foundation, and the Galápagos Conservancy. At a summit convened in 1995, conservationists from these bodies collectively determined that, because their grazing was destroying drip pools that were crucial to the habitats of native tortoise species, eradicating the goat populations on these islands was essential for preventing the tortoises' extinction. This program was early in a wave of invasive species eradication efforts, which continues today, with the goal of saving the islands' iconic native and endemic wildlife. More recent programs, including Project Floreana, target rats instead of goats and accordingly lack the same dramatic element of the goat eradication (like most rat extermination efforts, their primary tools are anticoagulant poisons, as opposed to helicopter-mounted rifles). Nonetheless, all these programs involve one species (humans) carrying out a systematic extermination of another, on behalf of yet others.

Project Isabela was a success. It used new, innovative methodologies to completely eradicate goats on the northern portion of the island. At least, almost completely. At the end of the program, the conservationists spared 266 animals they referred to as "Judas goats," and left them on the island for monitoring purposes (Galápagos Conservancy n.d.). These Judas goats had been a key part of conservationists' strategy for eradicating the rest of northern Isabela's goat population. Those involved in the effort used the fact that goats are social animals to their advantage; a number of the animals were given tracking devices that led sharpshooters to large congregations of goats. The riflemen would then shoot all of these animals, either from the ground or from one of the circling choppers used in the effort, sparing only the radio-collared "Judas goats" so that they could repeat this process anew (Campbell and Donlan 2005). When they were fitted with tracking devices, the Judas goats were also sterilized, so that allowing them to live would not compromise the broader eradication effort. Karl Campbell, a conservationist with a PhD in vertebrate pest management from Australia's University of Queensland, added another measure to make this Judas goat technique even more effective. In addition to sterilizing female Judas goats, he and his fellow conservationists gave them hormone implants that substantially increased the duration they were in heat (Campbell et al. 2005). This made the gregarious nature of goats even more advantageous, as male goats were especially attracted to these hormone-enhanced females.

This goat eradication on Isabela was a landmark moment for conservation in the Galápagos, and it had the intended effect of dramatically reviving the drip pools that are a vital portion of tortoise habitats. It inspired hope that other endemic species in the archipelago that are vulnerable to extinction might be saved if certain introduced species were removed in the same way. Based on common notions of environmentalism, though, the project makes for a somewhat unconventional gold standard for conservation practice. For one, as discussed in the previous chapter, conservation as we are used to imagining it tends to operate on the assumption that the best thing we, as humans, can do for nature is get out of the way. The basis of "fortress" conservation is the goal of sectioning off areas that are to be off-limits for development and other forms of deleterious human impact. Project Isabela, though, epitomized an approach to conservation that is quite a bit more hands-on. The systematized killing of a species, introduced or not, is undeniably a major human intervention into nonhuman nature.

The other aspect of the program that goes against the grain of many environmentalists' sensibilities is the unavoidable violence inherent to eradication. Of course, interspecies killing, including the use of firearms, is not necessarily antithetical to environmentalism. Many of the most vocal proponents of the early environmental movement in the United States were hunters whose mode of communing with the nature they hoped to protect involved killing some of it with firearms. In *Moral Entanglements*, Stefan Bargheer shows how the pastime of bird-watching in the UK shifted from collecting specimens with guns to capturing their images in the lenses of cameras or binoculars (Bargheer 2018). This involved a corresponding transformation of the underlying shared morality that structured the institutionalization of the hobby. For many modern environmentalists, the use of lethal force as a means of conservation is seemingly at odds with the value system that guides the goal of stewarding nature. Moreover, the eradication programs in Galápagos occur alongside other conservation efforts that also require immense investments of time, resources, and energy but are exclusively aimed at keeping species alive without involving the bloodshed of Project Isabela's goat pogrom. These include well-publicized and ongoing efforts to save various species of giant tortoise through the breeding of these animals at the Charles Darwin Research Center. Compared to projects like this, species eradication programs align less obviously with common cultural sensibilities of environmentalism and conservation that value the preservation of life; though the eradication programs ultimately strive to save native species, the systematic killing of animals is not, in and of itself, easily identified with that goal. As Hillary Angelo says in comparing the sensibilities of ornithologists and birders, "Killing animals you love is [hard] to comprehend because most of us are not killing animals anymore" (Angelo 2013: 351–52).

Likewise, killing animals *for* the animals you love is similarly hard to comprehend, as the violence and brutality of eradication work seem unbefitting of the compassion for the natural world that often motivates conservation.

Unsurprisingly, the conservationists behind these eradication programs have faced opposition from people and organizations who disapprove of the violence of their methods. Some of these detractors are animal rights activists who unapologetically value limiting the suffering of individual animals over more holistic notions of the health of ecosystems. Others have a different interpretation of environmentalism based on divergent conceptions of nature and what makes it valuable. This perspective is dubious of the stark lines drawn between native and invasive species and is therefore unwilling to protect the former at the cost of exterminating the latter. This chapter analyzes this moral controversy to the end of better comprehending the stakes of our current age of environmentalism. Ultimately, the way one decides whether species eradication programs like Projects Isabela and Floreana are justified and necessitated is an indication of the broader cultural values about nature and our relationship to it that structure an ecology of meaning. The rats on Floreana Island and the ethical dilemma of what should be done about them has implications that reverberate far and wide: Where do we, as humans, draw the line between good and bad nature, and how should we use the power at our disposal to shape the nature of the future?

This is a question made urgent by the climate crisis. As climate change becomes simultaneously more severe and more prominent as a public issue, there is a growing consensus that we currently live in a defining moment for our relationship to nature. But just as environmentalism has long encompassed concern over both the preservation of finite material resources and the conservation of sublime and beautiful landscapes or creatures, addressing climate change will require not only ensuring the basic conditions that make human life possible on Earth, but also ensuring the possibility that life on Earth can be joyful. In the abstract, the task before us can be and often is framed in the quantitative terms of emission reduction goals or preventing average global temperatures from rising above a given threshold. But practically, the radical reorganization of our economies, politics, and ways of life that accomplish those goals will require smaller-scale decisions that intentionally or unintentionally privilege certain forms of (human and nonhuman) life over others. Privileging endemic bird species over introduced rats on a single archipelago is not a decision that is crucial for ensuring humanity's survival, but it is emblematic of other, related questions. What kind of nature do we want to take into the future with us while we go about ensuring our own survival? How should those decisions be made, and who should make them?

The next section examines how interviewees involved in species eradication programs contend with the moral issue of killing for conservation and the attitudes they expressed about criticisms of their programs along these lines. In discussing the rationale behind species eradication, these conservationists often appealed to the "naturalness" of killing, drawing distinctions between the extermination of invasive species and everyday forms of interspecies predation. This perspective sheds light on the ecology of meaning that the animals involved in these programs are a part of. According to the different meanings they carry, animals can be divided into "natural" or "unnatural" hunters and deserving or undeserving victims of interspecies killing. This is further demonstrated by a mixed-methods content analysis of blog posts from conservation organizations, which visualizes these categories of natural and unnatural killing as they appear in these textual materials.

The rest of the chapter considers a possible solution to the moral problem of killing animals for conservation that may be posed by the use of synthetic biology for eradication purposes. These new approaches, which are built on recent advances in the field of genetic editing, could make species eradication possible without the violence of guns or poison. While doing so, however, they also may ratchet up the stakes of other moral questions at the heart of human relationships with nature: What is our role in the natural world, and what is an acceptable level of influence we can wield in order to save nature (before we start to compromise it)? By laying bare these broader implications, the possibilities of genetic editing for species eradication reveal the potent cultural tensions that lie beneath the moral question of killing for life.

THE MORAL QUESTIONS OF ERADICATION

In an interview with Carolina, an environmental lawyer working for Island Conservation, she emphasized the broad enthusiasm for Project Floreana among not only the conservationists involved, but the permanent population of the island as well. She noted that, after a slightly contentious beginning touched on in the last chapter, the conservation team had successfully worked with the community to make the project improve various aspects of the quality of life on the island. This effort included new structures like chicken coops that not only would assist in the poisoning program but would also be a long-term benefit to the island's agricultural production. She also emphasized, though, that the primary goals of the eradication program would benefit the population along these lines. Nonetheless, it was clear that the program still had its detractors. When I summed up what I was hearing, saying, "It sounds like no one really wants rats or feral cats on Floreana," she responded, "Well, I'm sure we'll hear from the people who do."

With this response, Carolina was referring not to the Floreana population, but to others who object to the wholesale killing of rats for conservation purposes. As she tells it, some critics of species eradication programs feel that they arbitrarily value the lives of native species over those of introduced ones. Brian, a conservation biologist with decades of species eradication experience, touched on similar themes when he expressed exasperation about his clashes over the years with advocacy groups including the Fund for Animals and PETA. In particular, Brian opined that "all the ... animal rights groups, what they just can't wrap their head around and ever accept is the death of individual animals."

Brian's run-ins with these animal rights groups have included one incident in which a letter decrying the violence of a mongoose eradication program he led reached his university president. As he tells it, the path toward these confrontations began after a moment early in his career that changed his perspective on the value of nature and the necessary actions to steward it and shaped his future expertise in species eradication. He tells me that, while completing his PhD research, he traveled to the California Channel Islands to study the social behavior of the goat population there. "The more I sat on hillsides watching goats," he told me, "the more I realized, 'Holy shit, this island is denuded.'" Quickly, his attitude about the animals he was there to study shifted: "[I realized] there's a much more important question to be asked and answers to be found with the notion of an invasive species affecting an entire ecosystem rather than all the cutesy wonderful things about social behavior."

This is the general ethos that has fueled his career since then. His work has been defined by developing methods to successfully eliminate animals on islands where he and others deem they are having a negative ecological impact. He has a way of describing this situation that puts animals like these feral goats in a special, dubious category that he calls "negative keystone species." This categorization describes a species that is not only out of place in an ecosystem, but has an anti-ecological effect in its new environs: "Well, if you think of a keystone species as an organism in a system that, if you remove it, what happens? Well, you have severe disruption and then there's a whole lot of shuffles and all kinds of different species searching for some new equilibrium. Well, if you take an invasive species and insert it into an ecosystem, what happens? Same damn thing. All kinds of shuffling and whatnot. Seeking some new equilibrium, except it's a very negative effect. Negative keystone species. It's the reciprocal of the keystone species."

For Brian, this calculus based on principles of biology and conservation science justifies and even necessitates killing species identified as having these damaging effects. During one such eradication program on California's San Clemente Island, prominent animal rights activists got wind of the plans to hunt goats and generated a controversy that temporarily prompted the US Navy, which has a base on the island and was leading the program, to pursue a trapping procedure instead of eradication. Brian was adamant that this approach was ultimately ineffective in reducing the ecological harm the goats caused: "They never even got close to catching goats in all the very deep incised, remote canyons on the edges of the island. They just got all these thousands of goats off the top plateau of the island." He went on to further dismiss the notion that more "humane" protocols might be available to assuage the moral discomfort some have with eradication work. Brian pointed out that trapping instead of eradication introduces the question of what to do with the animals once they are trapped, claiming that in the case of San Clemente Island's goats, "The trapper was caught in a livestock semi down to Mexico. He was going to unload all these goats. . . . They were going to be 'humanely' turned into tacos."[1]

Both Brian and Carolina view moral objections to eradication as myopic. For them, this perspective misses the bigger picture of the ecological greater good. After all, few would argue with the notion that a great deal of death and killing is intrinsic to the natural world. In fact, conservationists involved in these programs that I spoke with would often implicitly or explicitly justify their choice of methods using appeals to the killing inherent to nature. For instance, I asked Carolina about the pushback some of their work has gotten for its perceived brutality. She responded by indicating that she felt this perception stems from an unhealthy aversion to recognizing the inherent brutality of life in general: "Well, my perspective is . . . our generation and the last three generations are . . . not used to getting to the brutal part. . . . But life is brutal. If you give birth to someone, it's brutal. And now people just . . . have fears of that. And so you need to get back to 'What is natural?' . . . 'What do we need to do?'"

Carolina's perspective casts the work of eradication as not fundamentally different from the everyday natural processes that govern ecosystems. In fact, she views the systematic killing of invasive species as an extension of those processes. "It's nature," she told me. "In

---

[1] In a different part of our conversation, Brian noted that in another goat eradication campaign on a remote atoll in the Seychelles, the World Bank insisted on having a crew that accompanied the eradicators to salvage the meat of the goats for the people of that island nation. He told me he balked at this, telling them, "You can do that, but the cost is going to be greater than if you just simply left [the goats] . . . and bought these people steaks and had [them] air freighted from Europe." While this case was slightly different from the animal rights objectors he encountered elsewhere, it similarly emphasizes the tension between Brian's ecological calculus and others' notions of value and morality that might lead some to instinctively object to the mass killing of animals.

order to survive, you need to kill." Her particular interpretation of nature was further grounded in references to the behavior of invasive species: "That's what invasive species do. Rodents, in order to survive, they eat whatever they have in hand. And it doesn't matter if it's a turtle's egg, iguanas, or birds or whatever. . . . [But] people [say] 'you're killing something and killing is not right.'" It is this comparison, between the death caused by introduced species like rats and the killing carried out in eradication programs, that Brian implicitly calls attention to when he expresses frustration with animal rights groups' inability to "accept the death of individual animals."

The moral question of killing as a conservation tool is an especially potent topic in Ecuador. In 2008, the country adopted a new constitution that included a landmark stipulation that enshrined codified rights afforded to nature. This made Ecuador the first country in the world to enact legislation of this kind. Specifically, the four articles of chapter 7 of the new constitution pertain to the rights of nature or "Pacha mama," the earth goddess of the indigenous Quechua peoples. Article 71 enshrines "the right to integral respect for [nature's] existence and for the maintenance and regeneration of its life cycles, structure, functions and evolutionary processes." Article 72 declares that "nature has the right to be restored," and empowers the state to "establish the most effective mechanisms to achieve" this restoration in cases of "severe or permanent environmental impact." Article 73 attempts to prevent such impacts from occurring. This article is of particular relevance for the conservation activities in the Galápagos, as it decrees that "introduction of organisms and organic and inorganic material that might definitively alter the nation's genetic assets is forbidden." Finally, Article 74 pertains to the value of nature to humans, stating that "persons, communities, peoples, and nations shall have the right to benefit from the environment and the natural wealth enabling them to enjoy the good way of living."

In 2021, these landmark constitutional provisions were tested in the courts. The case centered on plans to mine gold and copper in Los Cedros, a protected forest area in the northwest of mainland Ecuador. The country's constitutional court ruled that these plans violated the rights of nature by threatening the wildlife of the forest, which includes the critically endangered brown-headed spider monkey

In this instance, the threat to endangered species and, consequently, to nature at large is, straightforwardly, human activity. However, a nonhuman third party, such as an introduced species, makes operationalizing the rights of nature more difficult. Carolina tells me she has heard from some Ecuadorians who believe that rats and feral cats, introduced or not, fall under the category of nature that should be protected by these constitutional provisions. For Carolina, this perspective is a perversion of these statutes' intention that misses the holistic picture of ecological health.

Nonetheless, the problem of introduced species makes conservation not simply a question of protecting the rights of nature, but also one of adjudicating between good and bad forms of nature. And if killing is indeed an intrinsic part of nature, then there are good and bad forms of killing as well. Where the line is drawn between good and bad nature, between good and bad killing, is a question of moral interpretation and meaning. For animal rights activists who would apply Ecuador's constitutional rights of nature to all nonhuman animals, including feral cats and rats, all killing is a violation of nature. For Island Conservation and other species eradication practitioners, certain forms of killing are not only justified, they are also natural.

The question of killing for conservation demonstrates how the symbolic meanings of individual species become broader moral cosmologies. In the negotiation of ecologies of meaning, we turn to individual animals as symbolic avatars of higher-order nature. How we assign categories of good or bad nature to these animals is how the moral values informing cultural ideas of nature are continuously constructed.

A Typology of Killers and Killed

The justifications for killing that species eradication practitioners turn to are grounded in a particular holistic cultural idea of nature. To analyze the role of the symbolism attributed to individual animal species (as well as meaningful categories like "invasive" versus "native," or "keystone" versus "negative keystone" species) for generating that working conception of nature, I turn to online blog posts published by various conservation organizations that work in the Galápagos.

These data are useful for this kind of analysis because they narrate the work that these organizations undertake in their own words, and thus offer a window into the underlying set of values they bring to it. For example, a 2021 post on Island Conservation's website reports on the situation in Gough Island in the south Atlantic Ocean, where introduced mice pose a threat to native bird species. The article is titled "Will Threatened Tristan Albatross Chick Survive after Invasive Mice Killed Its Mother?," and its first paragraph reads: "For the first time on record a Critically Endangered Tristan Albatross adult has been eaten alive by invasive non-native house mice. About a third of Tristan Albatross chicks are eaten by the introduced mice each year on Gough Island, a UK Overseas Territory island and World Heritage Site in the South Atlantic 2,600 km away from the nearest land mass of South Africa. Only two to three pairs of Tristan Albatross breed anywhere else on Earth. The mouse predation, and the threat of unsafe fishing practices, has placed them in danger of extinction."

This introduction gives us both a simple overview of the situation on Gough Island (the Tristan albatross may go extinct owing largely to the effects of introduced mice who prey on the birds' chicks and now, adult albatrosses as well). It also gives us a strong indication of the moral values underlying the ideas of nature that inform Island Conservation's work. While conservationists might explain the killing inherent to eradication work as an extension of the kind of killing that occurs in the nonhuman world, this passage makes clear that the particular predation being described is to be viewed both as not natural and as representing an urgent problem to be addressed. For one, the category of "invasive" applied to the mouse in question immediately places it in a taboo category of predator. Moreover, the act of predation itself is narrated with the phrase "eaten alive," which recalls a monster movie tagline.

According to the animals involved and the meaningful categories to which they are assigned, interspecies killing can take good and bad forms according to species eradication's ecology of meaning. If there are no humans or non-native animals involved, killing is an ordinary part of the animal world as governed by natural selection. Interspecies killing where a human or a rogue species they helped introduce kills a native animal, though, is an unacceptable form of killing that lacks the justification of naturalness. However, human extermination of an introduced species with the goal of preventing this bad killing falls into the good category.

To generate an empirical picture of these distinctions, I collected a sample of 1,263 blog posts from the websites of organizations involved in species eradication and conservation in the Galápagos (Island Conservation, the Charles Darwin Foundation, and the Galápagos Conservancy). Using automated Natural Language Processing methods, I then isolated the instances where killing of some sort is mentioned in the blog posts I collected. I then qualitatively coded each of these sentences for the species killing and the species being killed, as well as whether each was construed as invasive or native (humans were simply coded as "human").

In the initial round of coding, I recorded the species at the level of specificity used in the documents themselves. Rats, for example, were variously described as killing "birds," "seabirds," "albatrosses," "petrels," "kites," and myriad other avian species. In a second round of coding, I collapsed many of these into broader categories for the sake of interpretability. Each individual species of bird was recoded simply as "bird," rats became "rodents," and the same procedure collapsed several more specific entries into categories such as "reptile" and "crustacean." Because the conventions of language, popular knowledge and understandings of animal species, and formal biological taxonomy are imperfectly aligned, there were no universally applicable criteria to refer to in deciding what level of specificity to use in this round of coding. In general, I aimed to retain as much detail as possible while producing a coherent and interpretable network.

I took a conservative approach to coding species as native or invasive, wherein species were assumed to be native unless identified as introduced/invasive. Only instances of direct, intentional killing were included; descriptions of indirect effects of invasive species, such as the death of tortoises from habitat loss caused by the grazing of introduced goats, were excluded, for example. Some sentences identified by the automated text analysis yielded several pairs in this manual coding, while others yielded none. An illustrative example is the following sentence from a Charles Darwin Foundation blog post: "It is important to control or exterminate introduced species such as cats and rats so they do not continue to attack birds, and to maintain control of the workings of fisheries and vessels, to avoid incidental by-catch and other diseases." This sentence generated the following pairs of interspecies killing:

- feline (invasive) → bird (native)
- rodent (invasive) → bird (native)
- human (human) → cat (invasive)
- human (human) → rat (invasive)
- human (human → fish (native)

Both cats and rats are described as preying on birds, while the sentence also makes mention of human attempts to eradicate both of these introduced species. It also mentions the fishing industry, which kills fish either intentionally or as "incidental by-catch." Many of the sentences that contained no pairs mentioned the death of an animal from other factors, including the widely reported death by natural causes of Lonesome George, the roughly 101-year-old Pinta Island tortoise and last of his species.

From this coding I produced a network map (fig. 7.1) that provides a visualization of the role of interspecies killing in ecosystems, specifically as they are imagined in the public-facing press materials of groups working to conserve them. Pairs that are coded as "unnatural" (killing of native species by humans or introduced/invasive species) are represented in the bottom panel, while "natural" killing involving only native species or humans exterminating invasive species is shown in the top panel. The solidity of the arrows corresponds with the number of documents in which the pair appeared, where darker arrows indicate more frequent occurrences. The web of connections in the top (natural) panel is roughly as dense as the bottom (unnatural) panel, indicating a fairly even split between these two forms of killing as they appear in these documents. However, the "natural" panel is dominated by dark arrows originating with "human," which represent instances of human extermination of introduced species. Since many of these blog posts are specifically reporting on the activities of eradication programs, it is unsurprising that these connections would be so heavily represented.

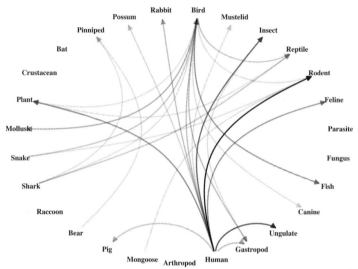
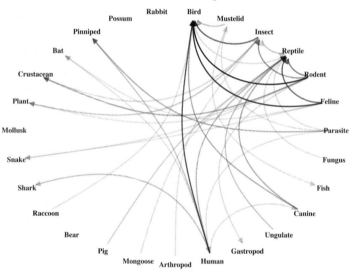

FIGURE 7.1 Network maps depicting the interspecies killing mentioned in blog posts of conservation organizations involved in species eradication efforts. Panels depict killing coded as "natural" (either humans eradicating invasive species or one native animal preying on another) or "unnatural" (any killing of native species not by another native species). Because the blogs discuss conservation efforts, human killing dominates both panels, especially the "natural" one.

Given this, the network map tells a different story when filtered to exclude pairs including humans. This second visualization (fig. 7.2) shows that, apart from human eradication efforts, invasive predation is far and away the most common type of interspecies killing that appears in these documents. The number of lines in the "unnatural" panel and their solidity indicates that this is true in terms of both the total count of pairs and their comparative frequency. The natural panel shows that other types of killing surely are still present in the data. Native birds prey on fish, snails, and other species in the blog posts, for instance. However, they much more frequently appear as the victims of invasive rodents, felines, and canines.

These network maps reveal an overarching narrative about nature that frames the use of species eradication for conservation. Interspecies killing may be an inherent feature of a "red of tooth and claw" natural world, but the killing that is emphasized most in these posts (the insidious predation of rats, cats, goats, and other invasive species) is cast as unnatural. This nuanced portrayal of the naturalness of killing breaks the animal kingdom into deserving and undeserving victims, portraying their deaths as accordingly tragic or routine. The blog post quoted above about the threat that mice pose to a population of albatross on a remote archipelago featured a gruesome photo of a bloody head wound one bird suffered at the hands of a rodent (it also had a warning at the top of the page advising the reader that the post included graphic content). This adds important texture to the appeals to interspecies killing that are used to justify species eradication. On one hand, these are attempts to cast these conservation methods as "natural." On the other hand, though, the justification hinges on the symbolic potency of birds and other native species as undeserving victims of predation, and likewise of rats and mice as killers that must be stopped.

Moreover, both sides of the debate about killing as a conservation tool are invested in the preservation of animals, but they arrive at polar-opposite prescriptions for how to go about this in practice. In fact, despite being mostly dismissive of or enraged by the perspective of animal rights advocates throughout our interview, Brian offered one indication that he understood where they were coming from. As he put it, "All these people, you know, PETA and Fund for Animals,

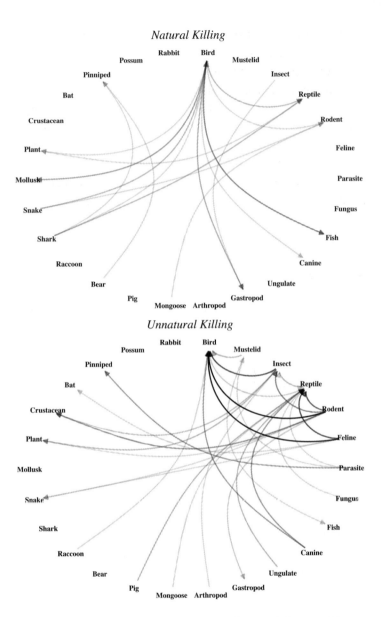

FIGURE 7.2 Filtered network map excluding pairs that contain humans. This new visualization suggests that, outside of eradication efforts, these blogs mention "natural" forms of killing substantially less than they do "unnatural" forms.

they are hopelessly ignorant and wrong for all the right reasons." After all, in the end, Brian wants to save animals too: just different animals and from a different threat to their lives.

What this comparison reveals is a difference in the kind of nature that animals symbolize for these divergent perspectives. Angelo (2013) argues that animals "make nature for us" at the level of the interactional encounter, and we might interpret the different perspectives around eradication in the same terms—as different cultural orientations to nature stemming from distinct experiences with nonhuman others. But in this case, the interactions in question between humans and animals are explicitly connected to overarching environmental values and must be interpreted in those terms. In this sense, it helps to consider these animals as cultural objects being deployed in different regimes of shared cultural meaning. Whereas for those involved in species eradication, a Galápagos tortoise symbolizes the cultural notion of evolutionary change as described in the previous chapter, for many others the value of nature is most clearly identified in instances of symbiosis and harmony. For them, the role of endemic animals as avatars of nature is fundamentally at odds with the use of violence to protect them.

Ultimately, debates about killing for conservation do not have answers grounded solely in rational scientific appeals. Rather, they hinge on different conceptions of what makes nature valuable and, relatedly, where lines are drawn between good and bad forms of nature. What if killing was not an issue, though? Considering this (not completely hypothetical) question further illustrates that these moral questions extend beyond squeamishness about or desensitization to the extermination of rats and other introduced species. Rather, that issue scratches the surface of the broader question of how and to what extent humans ought to intervene in an ecosystem in any way for the goal of preserving it.

A POSSIBLE FUTURE OF SPECIES ERADICATION

A vocal contingent of the conservationists involved in species eradication programs would welcome a way to make an end run around the issue of killing and its ethical ambiguities. A new proposed tool promises just that, but at the cost of introducing a host of other, new ethical considerations. A group of species eradication practitioners have been exploring the possibility of applying recent breakthroughs in the field of genetic editing to the goal of eliminating introduced rats on islands. Specifically, a tool known as a "synthetic gene drive" offers a way of ensuring the transference of a specific gene from one generation to the next. This could be used to proliferate a gene for infertility throughout a population of rats so that, within a few short generations, the animals would simply vanish, no massive deployment of poison needed.

This tool is based on CRISPR, which stands for "Clustered Regularly Interspaced Short Palindromic Repeats," a phenomenon where small pieces of genetic code are repeated regularly throughout a genome. These repeats exist in nature and were observed as early as 1987 in *E. coli* bacteria (Ishino et al. 1987), though the acronym CRISPR has in recent years come to be used as a shorthand in public discourse for the gene editing technologies developed based on them. Decades after they were first observed, the function of CRISPR sequences in bacteria was discovered to be an immunizing defense system against viruses (Brouns et al. 2008). Essentially, when a bacterium survives an attack from a virus, it stores a piece of its DNA in a CRISPR sequence and then, if attacked again, will use that identifying genetic code to destroy the attacking virus's DNA (Plumer et al. 2016). What makes this discovery a monumental one for the project of gene editing is that the immune system that functions through CRISPR is effective at cutting specific pieces of genetic code and replacing them with new code. Researchers quickly discovered that CRISPR sequences could be manipulated to target a designated genetic sequence and could be used to replace that sequence with a desired one. In 2013, scientists demonstrated that CRISPR could be used to modify the genomes of cells in mice (Cong et al. 2013) and humans (Mali et al. 2013), among other applications. The synthetic gene drive, or "mutagenic chain reaction" (Gantz and Bier 2015; Unckless et al. 2015), applies CRISPR beyond the level of individual organisms by ensuring that a given trait is passed on to the next generation.

Conservationists from Island Conservation are participating in a project called Genetic Biocontrol of Invasive Rodents (GBIRd) that is conducting research on the feasibility of replacing anticoagulant poisons with the gene drive in future eradication efforts. There are multiple reasons, some practical and others ethical, that many conservationists would welcome this approach over their current available methodologies. Practically, the switch from poison to genetic editing might be analogous to a switch from buckshot to a silver bullet. The current approach to eradicating rats comes with enormous risk of collateral damage. This is why, when I visited Floreana, crews were in the middle of constructing aviaries to house native bird species during the poisoning program in an attempt to avoid those animals inadvertently ingesting poison by preying on rats. If the gene drive worked as intended, it would not have this same risk, as its effects would be localized to the rat population.

The other reason the gene drive is an attractive option is it is seen by many people as a "humane" way to remove an introduced species. For those who sympathize with the conservation goals of species eradication but bristle at the thought of gunmen hunting goats from helicopters with rifles or rodents dying painful deaths brought on by anticoagulants, the promise of genetic engineering is especially alluring. Perhaps conservationists could save endangered species while simply sitting back and letting rats live out their remaining days in peace, eventually vanishing without incident.

The advent of CRISPR and the synthetic gene drive, along with the possible arrival of synthetic biology-based approaches to species eradication in the near future, introduces new wrinkles into the debate over killing as a conservation method that reveal the moral terms of that debate. What ethical trade-offs would this approach come with in return for less violently dispensing with invasive species?

CRISPR and Bioethics

Genetic editing has existed for much longer than the recent technologies based on CRISPR have. The reason these new methods are regarded as so revolutionary is less related to the things that they make possible that previously were not than to the fact that these applications, while technically possible before, can now be achieved more cheaply and efficiently. The same thing is true, in a sense, of the opportunities for species eradication presented by the synthetic gene drive. As any Judas goat left on an island after the rest of their caprine brethren were finished being gunned down would tell you, species eradication is clearly already possible. But if researchers confirm the viability of a gene drive approach, species eradication could become much more efficient and much less messy. While this promises to solve the moral problem of killing for conservation purposes, it pushes even deeper existing questions about the stewardship of nature to the foreground.

CRISPR and gene drives, as well as the field of synthetic biology more broadly, have been the subject of warnings from bioethicists anxious about the rapid progress of these developments and the need for careful consideration of their implications before they come into use (Billings, Hubbard, and Newman 1999; Cribbs and Perera 2017; Mulvihill et al. 2017). Possible uses of genetic engineering on human cells have prompted some to raise concerns over informed consent, or even the necessity of an "informed refusal" (Benjamin 2016a), with many claiming such interventions are unethical on the grounds that obtaining consent from affected future generations is simply impossible (Billings et al. 1999; Cribbs and Perera 2017).

Others have expressed anxiety over the potential for a lack of effective oversight over the use of this technology. While CRISPR has sometimes been heralded as a democratizing technology for its wide availability, Thomas Douglas and Julian Savulescu (2010) opine that the chief concern for ethicists regarding synthetic biology is the question of whether knowledge dissemination is inherently a social good, citing the potential for CRISPR to be used for biological terrorism or warfare. Sociologist of science and race Ruha Benjamin has warned that CRISPR, like any technological advancement, comes into existence within social conditions of hierarchy and inequality and promises to reproduce those dynamics in its application. Her analysis of these dangers inherent to CRISPR mostly deals with applications of the technology in humans. Whereas the utopian vision of gene editing is one in which public health problems are systematically dispatched with, Benjamin writes that genetic engineering technologies may reinforce "assumptions about which lives are worth living and which are worth 'editing' out of existence" (Benjamin 2016b). As she writes, "The bright line we may wish to draw between laudable and questionable uses of gene editing techniques is more porous than we realize." In other words, it might seem like common sense that the elimination of suffering caused by certain genetic conditions would be an obvious net positive, but there is a much hazier boundary than most acknowledge between this sort of application and a eugenics regime where the powerful few broadly impose their interpretation of genetic purity.

While these concerns over bioethics are the most prominent alarm bells ringing in the public discourse about CRISPR, there have been some notable warnings about the environmental risks of its applications in nonhumans. Typically, these have cautioned about the potential for unforeseen, downstream ecological impacts of using gene drives in animal populations (Esvelt et al. 2014; Rodriguez 2016). The National Academies of Sciences has issued a report to this effect, which warns that the development of gene editing technologies and the increasing viability of their applications are outpacing our ability to adequately assess their overall ecological impacts (2016). While this perspective is of vital importance, these environmental concerns have not been discussed with the same critical approach to power, inequality, and social structure that has been present in debates over bioethics.

Nonetheless, the potentially dangerous ethical ambiguities that Benjamin cites in human applications of genetic editing are also relevant for its application in nonhumans. As evidenced by the lack of universal consensus regarding the use of killing for conservation as well as what nonhuman species qualify for protection under the "rights of nature," it is not so simple to identify what constitutes good or bad nature. Even if many of the detractors for species eradication were won over by the less violent methodology made possible by the synthetic gene drive, the use of these technologies would usher in a new era where humans would have more power to act on the outcomes of these determinations. Whereas CRISPR uses in humans force us to think about what people, organizations, governments, or cultural worldviews will wield the power of this newfound influence over the gene pool, uses for conservation empower specific actors to dictate what the nature of the future will look like.

Many proponents of the use of the synthetic gene drive might reason that, on the relatively small scale of individual islands, these more philosophical concerns might be easily outweighed by the tool's conservation potential. In other words, a few empowered conservationists declaring rats to be bad nature solely in the context of an isolated, self-contained island, and, to use Benjamin's words, "editing them out of existence" there, would be worth making some detractors uneasy if it saved an endemic species that would otherwise die out. After all, rats live all over the world; will they really be missed if they disappear from a single island? However, the fact that rats have reached these remote locales, requiring species eradication efforts in the first place, is an indication that island ecosystems are not as easily sequestered from the broader global ecology as we might like to think. Should rats with the gene drive escape the "lab" of a remote island, there are fears that they could affect nontarget rat populations spanning an unpredictably large area (rats, after all, are famous for having followed humans to the ends of the earth). Such a development would mark a dramatic change in humanity's influence over the nonhuman world and would accordingly generate questions about what actors make decisions about the deployment of this new power. In this sense, the question of killing for life, of engineering the natural world, is an echo of a broader contemporary moment when the stakes of our relationship to nature, how we define it, and how we act on it are existential.

CRISPR and the Anthropocene

The recent breakthroughs in technologies based on CRISPR have been frequently described as precipitating a sea change in the field of genetic engineering and, thus, in humanity's relationship to genetics more broadly. The world of possibilities they open, from eliminating the suffering caused by certain genetically inherited medical conditions to saving endangered species (or even reviving long-extinct ones), implies a triumph of human ingenuity over the limitations imposed by nature. The depth of this possible transformation recalls a similar discourse on another fundamental shift in humanity's relationship to the natural world. In 2000, Paul Crutzen and Eugene

Stoermer published a newsletter for the International Geosphere-Biosphere Programme proposing that Earth had entered a new geological epoch, the "Anthropocene," wherein human activity is the dominant influence on global systems. Though it was not the first use of "Anthropocene," Crutzen and Stoermer's proposition popularized the term and prompted a surge of scholarly interest and research into the idea. The argument for the Anthropocene is based on the principle that "what matters when dividing geological-scale time is global-scale changes to Earth's status" (Lewis and Maslin 2015: 171). Given the effects of carbon emissions, the transformation of landscapes for agriculture, and many other industrial human projects, the Anthropocene would seem to qualify, though the formal criteria necessary for the demarcation of a geological epoch (records of major shifts in stratigraphic material) constrain the viability of arguments for a specific start date of the era (Lewis and Maslin 2015). Several such origins have been suggested, ranging from the advent of agriculture, to the Industrial Revolution, to the first detonation of an atomic bomb with the Trinity test in 1945 (Lewis and Maslin 2015).

From a cultural perspective, the scientific viability of any of these individual start dates may be less important than the conversations around them that attempt to grapple with the notion that humanity's relationship to the nonhuman world has fundamentally changed. In this sense, the importance of the term is as much linked to the social changes it describes as it is to the corresponding geologic ones. As Dipesh Chakrabarty phrases it, the notion of the Anthropocene "[spells] the collapse of the age-old humanist distinction between natural history and human history" (2009). In this sense, the Anthropocene and CRISPR both force us to consider a relationship between society and nature that transcends notions of a fundamental separation between those two realms. They are parallel sea changes to this relationship, where the terms of one's influence over the other are dramatically overhauled. In the macro sense, where does nature truly exist, when the Anthropocene leaves not even the most remote and vast ecosystems beyond the scope of human influence? And where is nature in the much smaller sense, when our bodies and those of nonhuman animals are no longer solely the products of extra-human processes, but may be created in part by synthetic biologies?

Moreover, the collision of these two changes compels us to reconsider how micro and macro levels of the human-nature relationship are linked. Anthropologist Margaret Lock has sounded alarms within her discipline, arguing that revelations from epigeneticists that the human genome is reactive to different environments demand a retheorization of "the body" and a reevaluation of the "nature/nurture debate" (Lock 2013a, 2013b, 2015, 2017). She situates these developments within the Anthropocene, suggesting a connection worthy of scholarly attention between the human condition on the most microscopic of scales, our genetic makeup, and the age of ecological crisis we live in (Lock 2017). When our day-to-day lives might often seem abstracted from a global crisis like climate change, Lock challenges the distinction between these scales, hinting at how we might feel the Anthropocene in our bones. Meanwhile, the advent of CRISPR technology promises to transcend the micro/macro distinction in a corollary way. The proposition of using gene drives as tools for species eradication indicates the powerful potential of the technology's effects to go beyond manipulating cells and bodies to transform landscapes and ecosystems. The fears that those effects could unintentionally spread far beyond their area of application further demonstrates that there may be no isolating the micro from the macro. Whereas the Anthropocene's effects may extend into the genomes of organisms, CRISPR technology promises to impact the genomes of ecosystems.

In the context of species eradication programs, these implications raise further questions where cultural meaning is concerned, especially in the Galápagos. Endemic species are valued in part based on scientific principles like biodiversity. However, even these seemingly rational, scientific determinations are rooted in cultural values. As discussed previously, the wildlife in the Galápagos is iconic not simply because it is native, but also because of its deep cultural connections to the theory of evolution. The possibility of intervening at the level of genetics to protect an endangered species might compromise some of that symbolic value by disrupting the terms of evolution as we know them. If the reverberating, interconnected ramifications of the Anthropocene and genetic editing mean that ecosystems are human influenced at both micro- and macroscopic levels, can a Galápagos tortoise still function as the symbolic beacon of natural selection it has become?

To phrase this a different way, the rapidly and dramatically shifting terrain of our contemporary cultural and environmental moment, in which humans are reckoning with the global impact of anthropogenic processes while also devising new ways to transform the natural world at microscopic levels, represents an upheaval in the relationship between ecology and ecology of meaning. The moral questions explored in this chapter—whether killing is justified to save endangered species, whether it is our place to engineer the genomes of species to right historical ecological wrongs—hinge on this relationship. The cultural meanings of animal species, from invasive rats to endemic birds and tortoises, comprise and exist within the comprehensive whole of larger cultural ideas of nature. But the rat is not solely symbolic any more than it is solely material. Rather, the symbolic weight of these animals is contingent in part on their material reality. Likewise, what nature means to us is connected to the material reality of the ecologies that surround us. The balance of this relationship is liable to be affected by any alterations we make to the materiality of ecosystems.

The magnitude of the age of CRISPR and the Anthropocene means we are simultaneously struggling to preserve a livable planet and deciding how we will leverage our only increasing capacity to influence natural processes. In this context, the seemingly localized issue of species eradication on individual islands is a window into the much wider questions of cultural meaning and nature. For good reason, our current ecological crises are most often framed as existential threats—we can either summon the collective will to stop climate change from making the planet uninhabitable or face the end of humanity. But imagining and charting livable futures will involve negotiating what nature means and how we value it, in addition to simply preventing global temperatures from breaching a given threshold. Rebecca Elliott has evocatively argued that making sociological sense of the climate crisis necessitates a "sociology of loss" (Elliott 2018) that attends to "what does, will, or must disappear," in contrast to the more widespread framework of "sustainability." Reorienting our perspective on the climate crisis in this way emphasizes the multiplicity of possible environmental futures that transcend the simplistic binary options implied by the lingering question "Will we save the planet?" As the moral controversy around species eradication points to, the abundance of possible outcomes also means an array of decisions on where to draw the line between good and bad nature.

Moreover, the questions of cultural meaning that lurk in our collective grappling with an age of ecological crisis are unlikely to be addressed by anything close to a universal consensus. Competing ideas of nature, what makes it valuable, and how best to protect it will be reconciled unevenly by actors with the levers of various instruments of power (from gene drive technologies to geo-engineering projects to national economies and supply chains) at their disposal. In this light, the notion of genetically erasing rats from an island or exterminating them with poison is a microcosm of our continual reimagination and reconfiguration of the natural world and our place in it.

× **EIGHT** ×

CONCLUSION

Among the many research trips I made and interviews I conducted for this project, with government bureaucrats on the tenth story of a Los Angeles high-rise, pest control officers in the frigid prairie of western Canada, vacationers on an island just south of the equator, and more, one moment from a particular conversation sticks out as particularly emblematic of the story of rat extermination writ large. The interview was with Brian, the conservation biologist who pioneered species eradication techniques and who was mentioned in the previous two chapters.

A small digression for a bit of background is warranted here. While Brian and I spent most of our time recounting the history of his long career, which took him to islands across the globe for the purpose of devising protocols for eliminating invasive species, at one point we turned to the future of rat eradication. We discussed how CRISPR-Cas9 has made gene editing easier and more affordable than ever and has also paved the way for the "gene drive," the technology that ensures the transference of genetically edited traits across generations. For now, the work of eradicating rats on islands like Floreana in the Galápagos involves a massive program of baiting and poisoning using anticoagulants. As one interviewee from the organization Island Conservation summed it up, "Those are the tools that are available to us now." However, as discussed in chapter 7, the scientific advances in genetic editing could dramatically overhaul these efforts. Theoretically, if conservationists harnessed the gene drive, they could, with surgical precision, render a population of target animals unable to reproduce after a few generations (which would elapse quickly in the case of rats). The remaining rats would then die off with little drama, the eradication having essentially taken care of itself.

I wanted to get Brian's thoughts on this potential future, given his long career in species eradication and his fervent, unapologetic embrace of the violent methods it currently involves for the greater good of saving native species. Given the controversies surrounding this violence inherent to species eradication explored in the previous chapter, many conservationists and others who support the goals of eradication programs are excited about the promise of the gene drive as an end run around these moral quandaries. Moreover, for all the exasperation he vented during our conversation about animal rights groups and their aversion to killing for conservation, Brian still at times expressed a sensitivity to the painful deaths to which the targets of eradication efforts are subject. For instance, at one point he described a method of rat extermination that is used less often than trapping. Instead, this method involves coating a small pipe or tube with poison-laced petroleum jelly. Because rats can seldom resist walking through small tunnels like these, he told me, the rats become coated in the poisonous goo and invariably ingest it while trying to clean themselves. "Very few people have ever dared do it," Brian said, "because of the bad vibes people get from it."

It was this discussion of the gruesome aspects inherent to rat extermination as currently performed that prompted me to get his opinion on the gene drive. Indeed, Brian told me he viewed it as a more humane option compared to the painful deaths brought on by rat poisons and that he was enthusiastic about its potential, sentiments that were shared by other key players in the eradication programs in the Galápagos. "I think it's a great idea" was his immediate response. As discussed in chapter 7, though, many see the use of genetic editing for conservation as posing even more serious ethical dilemmas. Some such concerns take issue with the premise of these approaches themselves, finding serious bioethical problems with large-scale genetic editing and the extent to which it would allow certain actors to "play god." When I spoke with Brian, though, I brought up a more practical concern that sets aside these more fundamental questions: While the theory behind using the gene drive assumes a self-contained animal population, what if one of the genetically modified animals stowed away on a boat or otherwise escaped the "laboratory" of the island (after all, this is how many of the species that are targets for eradication arrived at island ecosystems in the first place)? Would the gene drive spread out of control? Would an entire continent's (or the world's) population of rats rapidly vanish?

Brian was nonplussed by these hypotheticals. "Do you think anybody would really care?" he replied.

This question Brian posed in response to my own question feels like it lies beneath the surface of this entire examination of the global project of rat extermination. The real work of rat extermination is not a glamorous pastime. In most cases, it involves dank, dirty places, toxic chemicals, and, if successful, rodent corpses in need of proper disposal. But with that unpleasantness not a part of the equation, how many people would be bothered by a rat-free world? I am not qualified to speculate about the downstream ecological effects that might come from rapidly erasing rats from the Americas or from the world, but the fact that a trained biological conservationist posed this question made me wonder whether that contingent of rat defenders could succeed in persuading many others to come over to their side. As things are, few people are vocal in defense of the rats we currently exterminate. Moreover, the context in which Brian posed the question reminds us that it could be more than a simple thought experiment in the not-so-distant future. Technologies like CRISPR-Cas9 promise to give those who wield them unprecedented power to purposefully shape the material landscape of the world. Meanwhile, the urgency of the climate crisis demands that we exert control over the world in a different way. As we, through our unequal systems of power and authority that are themselves the object of contest, imagine and enact future worlds, what will we do with the rats of the world?

As I highlight this looming question, my goal is not to provide specific policy recommendations or a moral judgment on the rightful fate of rats. Having examined these three cases to the end of understanding cultural meaning as opposed to adjudicating this moral question, I maintain a somewhat ambivalent attitude about the morality of rat extermination (and animal extermination more broadly). I am sympathetic to the motivations of each rat extermination project considered in this book. Protecting the livelihood of family farms in Alberta is a worthy goal. When the COVID-19 pandemic upended our lives, nearly everyone became an amateur epidemiologist, suddenly paying more attention than ever to the latest research on the factors that shape the transmission of infectious diseases. With this recent context, it is hard not to be supportive of public health measures like those aimed at preventing the spread of typhus in downtown Los Angeles. Meanwhile, when I traveled to the Galápagos Islands as a researcher interested in the constructed nature of categories like "native" and "invasive" species, I was nonetheless mesmerized by the tortoises, mockingbirds, iguanas, and penguins I encountered during my stay in much the same way so many other visitors to the archipelago are. In between interviews and observations, I joyfully snapped the obligatory photographs of these creatures without which seemingly no visit to the islands is complete. Setting aside the cold, calculating logic of ecological science, I would much prefer that those animals not die out if that fate could be avoided.

On the other hand, the violence and killing that is unavoidable in the work of extermination is difficult to swallow, these compelling rationales notwithstanding. And while I am sympathetic to the motivations for rat extermination, I am also happy that it is not me who has to carry out the extermination itself. More pressing, at least from the perspective of cultural meaning, than whether or not we collectively decide that exterminating rats is important and justified, though, is recognizing that we make such determinations based on socially determined priorities, rather than objective scientific truths. In fact, the extent to which we are bothered by the killing of nonhuman others is highly contextual. Most people would never feel remorse about killing a mosquito, or even eradicating whole species of them as a public health measure, though they might find monarch butterflies or other flying insects beautiful. We might marvel at song birds through our windows but forgive our pet cats when they prey on those same birds and leave their corpses at our doors. Many American children have probably happily held out feed pellets for sheep at a petting zoo before eating lamb chops for dinner. Moreover, the social context that informs our toleration of killing in these various contexts is the product of a meaning-making process that carries the baggage of unequal power systems. Put another way, programs like the ones analyzed in this book are undertaken by governments, organizations, and private companies that are shaped by the social inequalities that surround them. The implications of the work they do, both material and symbolic, are not immune from these inequalities either. It is these points I want to highlight by returning to the two overarching arguments outlined in the introduction.

First, rat control is a social practice that draws and clarifies the boundaries of nature and society.

As argued in part I, animals are uniquely positioned to participate in the process of "boundary work" and rat control is an exercise in clarifying several meaningful boundaries, both spatial and symbolic. In Alberta, these include the geographic border between Alberta and Saskatchewan, and the discursive contours of Albertan collective identity. Controlling the lives and deaths of rats lends order to these distinctions. Importantly, though, these boundaries are expressions of the underlying notion of separate spheres for nature and society. What makes Alberta's claim to rat-free fame a moral accomplishment is the fact that, through this status, the province maintains a preferred balance between the human social world and the nonhuman world in its proximity. As the rhetoric of Alberta premier Ernest Manning during the 1950s attests, Alberta's cultural identity is closely tied to agriculture and deeply valued notions of hard work and determination that are ascribed to that industry. As narrated in Manning's addresses marking the province's Golden Jubilee, Albertan identity was forged through the trial of western Canada's harsh wilderness and is embodied by the spirit of idolized pioneer settlers who persevered through blizzards and dust bowls to lay the foundations for the province. The notion of a rat-free province is an expression of that hardy spirit's triumph over the natural elements. In contemporary Alberta and nearby Saskatchewan, the sentiment that having rats is an indictment of one's integrity as a farmer still resonates and even occasionally hampers collective rat control efforts along the border.

The concept of the "indoors/outdoors divide" introduced in part II and its importance for urban nature further illustrates how rat control attends to our cultural relationships with the nonhuman world. Cities have a complex relationship to nature. As massive agglomerations of human enterprise, they are paradigmatically viewed as concentrations of artificiality that are antithetical to nature. Yet nature abounds in cities in many different forms, and scholars like Angelo (2021) and Bell (2018) have convincingly argued that cultural ideas of nature are central to visions of desirable urban life. This complex relationship lends an ambivalence to city officials' practical efforts at managing their cities' relationship to nature. The nonhuman world must be folded into the fabric of urban life on specific terms, where "good nature" and its benefits are available to residents while they remain protected from the dangers of "bad" nature. These goals, which often appear to be in conflict with each other, are accomplished both materially and symbolically by managing the indoors/outdoors divide. In this context, rat control functions to make this boundary both significant and secure. While ultimately both the nature/society divide and the indoors/outdoors divide are concepts invested with shared cultural meaning more than they are material realities, they nonetheless are important for shaping social experiences of the city. While rats in general occupy a liminal space on the border between nature and society by thriving off the human excess of cities, the rats in LA's City Hall served as a reminder of the permeability of the indoors/outdoors divide. Moreover, they did not pose only a symbolic threat by breaching the walls of high-rises and forcing office workers to come face-to-face with the multispecies ecology that LA supports. By acting as a vector for typhus, their presence was a material threat as well. The response to this rat problem thus had to act in both material ways (making it more difficult for rats to take up residence in City Hall) and symbolic ones (guarding public faith in the indoors/outdoors divide, permeable as it may actually be). This, in a broader sense, is the work of keeping bad nature at bay.

Finally, the species eradication efforts in the Galápagos Islands explored in part III grapple with the question of who or what can legitimately affect nature for the better, as well as the underlying question of what qualifies as nature itself. The very goal of species eradication is to protect a preferred notion of what nature is, what it looks like, and what forms of life it allows. While the decision-making process that results in a species eradication campaign draws on scientific reasoning and expert input from biologists, killing one species to save another is nonetheless an expression of cultural values. Those values are embodied in and negotiated in reference to the individual animals who make up material ecologies and ecologies of meaning. In other words, when conservationists decide to exterminate a species to save another one, they affirm a specific categorization of the nonhuman world into good and bad nature. In the Galápagos, an omnipresent factor in this determination is the mythologized cultural legacy of Charles Darwin and the theory of evolution. While individual organisms like giant tortoises have been transformed into potent avatars of the laws of nature themselves, rats, as proxies for human influence that upend ecological conditions through habitat destruction, compromise the resonance of those other animals as cultural objects. Eradication thus intervenes in both the material ecology and the ecology of meaning by attempting to actualize an overarching cultural idea of nature.

Second, rat control enforces an implicit hierarchy of living things that mirrors and is entangled with social inequalities.

While rat control manages boundaries as described above, in doing so it necessarily wades into the discursive space of power inequalities where it occurs. Put another way, the border between nature and society that rat control negotiates is far from a value-neutral distinction. David Pellow's concept of "social discourses of animality" (2017) describes how references to the nonhuman world are used to police acceptable human behavior and to reaffirm social inequalities. That notion rests on the assumption of a symbolic hierarchy of living things that orders various forms of nonhuman life in their proper respective places below humans. Likewise, rat control is premised on the fundamental idea that rats are expendable, killable creatures.

In part I, we saw how being "rat-free" became a resonant narrative for a province invested in asserting its regional identity in opposition to the rest of Canada. However, the idea of Alberta as a rat-free island cut out of the vast expanse of otherwise rat-inhabited North America did not take hold merely as a regional eccentricity that set it apart from the other Canadian provinces purely arbitrarily. As illustrated by the promotional materials produced in the early years of the rat control program, making Alberta rat-free meant eliminating a symbolic villain that was framed as set on compromising Albertans' goals, values, and livelihood. This was accomplished not just by informing farmers on the Alberta prairie that rats could be a detriment to their agricultural production, but also with symbolic flourishes like positioning rats in posters as looming invaders along the Alberta-Saskatchewan border. Those same posters warned "Rats are coming!" or commanded residents to "Kill him!" The fact that those posters were produced by the Department of Public Health's Division of Entomology subtly reinforces the notion that rats are killable creatures at the bottom of a symbolic hierarchy.

This narrative of Alberta's anti-rat campaign found resonance in a province where various other forms of nativism have gained prominence and shaped social inequalities. Whereas a rat-free Alberta represented a certain notion of moral purity, the provincial government that oversaw the beginning of the rat control program was also concerned about other perceived threats to the province's collective moral character. The effort to eliminate rats shared a symmetry with the Manning government's moralized opposition to communism, whether in the external form of a nuclear-armed Soviet bloc or in the internal form of organized labor. Meanwhile, Alberta enforced notions of moral purity in another way, through a forced sterilization program that disproportionately targeted various already marginalized populations (Grekul et al. 2004). The symbolic hierarchy of species on which rat control rests in Alberta and the prominence of nativism there are mutually reinforcing. An illustration of that principle came in 2022, when the decades-old rat program provided a ready-made metaphor for a Calgary white supremacist organization's racist rhetoric.

In Los Angeles, the goal of bringing good nature into the city center while managing the threats of bad nature implicitly requires populating those categories with various elements of the nonhuman world. City officials must balance amenities like street trees, public parks, and the aesthetic beauty of the biophysical landscape against dangers like vulnerability to natural disasters and public health threats like infectious disease. Los Angeles is all too familiar with this give and take. Its setting, where the purple-toned ridgeline of the San

Gabriel Mountains meets the picturesque coastline of the Pacific Ocean, offers ample forms of good nature, but Angelenos are vulnerable to what Mike Davis (1998) termed a "dialectic of ordinary disaster" where floods, wildfires, and earthquakes turn LA's material landscape deadly. Los Angeles famously encased its namesake river in concrete to stop what otherwise might be a key form of good nature from overflowing unpredictably (Price 2006). While the categorization of good and bad nature is tied to these material realities, the meaning-making process that assigns given forms of nature to either bucket may outpace that materiality. Chapter 4 showed how rats came to carry symbolic associations with disease and dirtiness that transcended the role they played in spreading typhus (especially in the context of previous outbreaks that were more closely tied to other, less villainized vectors like opossums and domestic pets). While the rat control effort in City Hall functioned as an epidemiological measure, it also reinforced rats' lowly place on the symbolic hierarchy of species among the most killable forms of bad nature.

The inseparable connections between the saga of City Hall's rat problem and Los Angeles's ongoing struggles with homelessness and housing insecurity further reveal how social inequalities and interspecies hierarchies become linked. City responses to both the rat problem and homeless encampments fall under the joint project of managing the material and symbolic resonance of the indoors/outdoors divide. By inciting fears over typhus in city employees working inside the Civic Center, rats bring dangerous bad nature to the indoors. Residents of homeless encampments inversely breach that barrier by living and storing personal possessions on the sidewalk. In this context, city management of these two problems necessarily bleeds into the contentious and unequal politics of access to public and private spaces in the city. Based on my interviews, I am confident that many if not most of the individual people engaged in the city's tandem efforts to combat the rat problem and address its homelessness crisis have real empathy for those who sleep on the street in downtown LA and are in favor of progressive measures to address the city's ever-worsening housing crisis. The slippage that sometimes occurred in our conversations, though, where the "rat problem" and the "homeless problem" became nearly interchangeable, sheds light on another instance of rats becoming symbolic proxies for the most marginalized of human society. Moreover, the notion that the encampments were the primary precipitating factor in the rat infestation cast unhoused individuals as the cause of a typhus outbreak instead of the demographic most victimized by it.[1]

Finally, in the Galápagos Islands, we get a window into the connection between these two overarching arguments and their broader implications. If rat eradication is an exercise in delineating nature from society, in separating good nature from bad nature, then a crucial question that follows is who, in the broadest sense, is deciding what nature is and how to enact that vision. The pursuit of environmental conservation is a practice of enacting a material vision of environmental values. The questions of what makes nature good, what kinds of nonhuman life are worth saving, and what kinds of tools and methodologies we ought to use to do so are all shaped by underlying cultural ideas about our relationship, as humans, to the nonhuman world. Those ideas are not static or singular but are instead dynamic and contested. As discourse over the application of Ecuador's landmark codification of the rights of nature illustrates, there are multiple, competing notions of where the line between good and bad nature is. Through the concept of the ecology of meaning, we can see how these competing holistic ideas of nature are expressed in the meaning we ascribe to individual animals as cultural objects. Some detractors object to any and all species eradication programs on the grounds that they arbitrarily apply the rights of nature or that the killing of animals itself is simply wrong. Nonetheless, the conservation programs operating in the culturally charged landscape of the Galápagos Islands offer an exemplary case for how conservationists can avoid imposing a particular valuation of nature with a veneer of scientific truth providing a justification. Project Floreana overcame a rocky start to the relationship between conservationists and the islands' permanent human population to work collaboratively with

---

1   This perverse narrative framing highlights Steven Thrasher's (2022) argument regarding the mutually reinforcing nature of material inequalities, stigma, and infectious disease.

those most direct of stakeholders and craft a program that aligns with their interests and values.

The importance of democratic and just approaches to the management of nature and the difficulty of achieving them are only increasing as shifting technical and biophysical terrains steadily upend the relationship between humans and the nonhuman world. The Anthropocene troubles notions of a nature/society divide more than ever before and poses new wrinkles in the negotiation of ecologies of meaning. In terms of the Galápagos, for example, can giant tortoises and Galápagos finches perform their symbolic work of embodying natural selection if the terms of evolution are everywhere altered by various forms of human activity and influence? How do we value nature once we let go of the notion that it ought to be fundamentally separate from us? Moreover, as technologies like CRISPR-Cas9 and the gene drive expand our tool kit for altering the natural world, what measures will we take to enact our visions of nature, and who will make those decisions?

My hope is that the implications of these two overarching arguments will prompt future analysts who examine the intersection of culture and the environment to pay greater attention to the role of "bad" nature for environmentalism. Some biologists and conservation scientists have begun moving in this direction by discussing ecosystem "disservices" alongside "services" (Lyytimäki 2014; Vaz et al. 2017). Frameworks like these challenge the implicit assumption embedded in concepts like "ecosystem services" that nature is fundamentally a boon to human society. This notion constrains our imagination of possible environmental futures. While the work on ecosystem disservices is a welcome development, duly considering bad nature will require critically evaluating the end goals of conservation from a standpoint of cultural meaning. In other words, creating a future that staves off the worst forms of ecological collapse will necessarily mean not only protecting the nature we interpret as imperiled, but also eliminating other forms of nature that do not fit those imagined futures: the "rats" of the world. Where we draw the line between good and bad nature is not a foregone conclusion, but is rather a site of contestation that mirrors social inequalities and struggles for power.

In that vein, the note on which I want to conclude this book is not that all killing of animals by humans is inherently wrong or unjustifiable, but that the decision to kill nonhuman others, especially when it is in service of some broader socially important goal, is a culturally significant choice with broader implications than meet the eye. There is no objective formula to decide when it is justified, either. When we create the worlds of tomorrow, we will make decisions that will necessarily affect other forms of life either positively or negatively. If we decide saving particular species on tropical island ecosystems is important, some rats will have to die. How we arrive at that decision (and who the "we" in question are) is as important as the decision itself. Just as the steady accumulation of carbon pumped into the atmosphere boomerangs around in the material forms of superstorms and rising global temperatures, so too does the cultural meaning of our interactions with nonhuman others have a way of echoing back through our social worlds. The principle of climate justice ought to compel us to tackle the climate crisis while addressing the historical inequalities that leave some populations much more vulnerable to rising sea levels and other environmental maladies that they share disproportionately less responsibility for causing. Likewise, this same focus on justice and inequalities necessarily must be applied to the imagination of nature to be saved and the implications of how we save it.

ACKNOWLEDGMENTS

For the past several years, whenever someone I'm meeting for the first time has asked me about what I study as a sociologist, I have instinctively braced myself before answering, knowing that my response might be met with some confusion. "My book project is about rat control," I'll typically say, before explaining what's so interesting to me about the topic and hopefully bringing whomever I'm talking to around to the idea that it totally makes sense for a sociologist to be studying rat control. All of this is to say, this project has relied on other people who saw what I saw in it, which was not always a given. It's hard to express exactly how fortunate I feel that one of those people was Elizabeth Branch Dyson, who took a chance on the book and then seamlessly shepherded it through the editorial process at University of Chicago Press.

Of course, by the time Elizabeth heard about the project, I had had some practice explaining the idea behind it. The first big hurdle was bringing an earlier version to my PhD advisers, David Pellow and John Foran. With this in mind, I need to thank David and John for all the typical reasons one thanks their advisers (the close readings and generous feedback they provided over the years, the careful guidance crafting the research, the encouragement and endless positivity when I hit bumps in the road, and on and on), but what sticks out the most when I think about their mentorship is their reactions when I first pitched them a somewhat eccentric project about rat extermination. I was nervous then that they might think the idea was a bit too "out there" for sociology research, and in retrospect I appreciate how reasonable it would have been for one or both of them to advise me to do something a little more mainstream. On the other hand, I also realize now that, knowing them, there was little chance I was going to get anything other than the enthusiastic support I got from them at the time. I'm incredibly grateful for how they encouraged me to follow my instincts, and for how they saw sociological value in what some might have dismissed as a wacky idea. I'm especially indebted to David for helping me think through analytical challenges and brainstorm theoretical ideas on countless walks through downtown Santa Barbara.

I've also been lucky to have so many more mentors, formal and informal, that helped shape this book. Simonetta Falasca-Zamponi inspired me to study culture and provided a critical eye that pushed me to make everything I wrote that much clearer and that much more convincing. Kum-Kum Bhavnani gave me a methodological tool kit for qualitative research and let me ride her coattails to the Galápagos Islands. Peter Alagona helped me expand my interdisciplinary voice. Jon Cruz posed new, generative theoretical questions even when what I thought I needed was clarity. Hannah Wohl read countless drafts of pieces of this project and helped brainstorm ways to give amorphous ideas shape and life. And John Mohr gave me a template for how to bring creativity to the measurement of cultural meaning, and I wish I had known to ask all the questions I have now while he was still with us.

While researching and writing this book, I had the pleasure of collaborating on numerous other research projects. The book itself benefited enormously from these projects, which broadened my perspective and compelled me to rethink things from new angles. My coauthors on these projects also left their marks in ways that transcended the intellectual labor of sociological research, though, as friends and supportive colleagues. Zack King and I have been applying our sociological imaginations to politics and culture since we were both in high school, long before either of us first encountered that term in an introduction to sociology course and became inspired to pursue sociology PhD programs. Neil Dryden, who taught that introductory sociology class, has since become a dear friend, collaborator, and instigator who insists on pushing the boundaries of the discipline to be both more creative and more fun. In addition to coauthoring a paper with me, Adam Davis nearly single-handedly taught me how to code in R in between anchoring our pub quiz team and commiserating about the San Francisco Giants. After we bonded over the shared experience of researching "uncharismatic" animals, Caleb Scoville became an endless source of intellectual exchange, needed feedback, and professional support. A chance email exchange with Hesu Yoon turned into an incredibly fun collaboration on a paper that expanded methodological and theoretical horizons for both of us.

It's no secret that the landscape is a bleak one these days for recent or impending PhDs hoping to make a career in academia, especially in the humanities and social sciences. Accomplishing that goal has never felt like an inevitability for me, especially when I have seen so many incredibly talented friends and colleagues pushed out simply because of a bad job market and the collisions of other structural and personal factors. That I've made it as far as I have has been thanks to both the whims of random chance and the people who have done whatever was within their power to help my odds. In addition to everyone mentioned above, these include Hillary Angelo, Clayton Childress, Terrence McDonnell, Neil Gong, Fernando Domínguez Rubio, Nick Wilson, and others who provided me with mentorship, letters of recommendation, generous feedback, and other forms of support, often despite my being a graduate student or postdoc at a different institution. Heather Lynch and the Institute for Advanced Computational Studies at Stony Brook University threw me a lifeline when it looked likely that I might have to abandon hopes of continuing on as an academic and finishing this book.

More than anything else, I've been buoyed by my friends and family the past several years. Doing a PhD at my hometown university meant always having support from my mom and brother, Max. Having family close by made the process bearable, especially when the world shut down for the COVID-19 pandemic. Likewise, authoring this book would not have been possible without friends like Indiana Laub, Steve Stormoen, Rachael Rhys, Diane Phan, Brian Tyrrell, Kevin Brown, Tim Duggan, Andrea McKenna, Brian Lovato, and Daav Feldman.

More than anyone, I have my partner, Laura, to thank as this book reaches print. She convinced me in the first place that writing a book was a tangible goal I could achieve and not just a distant hypothetical to dream on. It wasn't until after she offhandedly suggested that I should put together a book proposal one day that doing so suddenly stopped seeming like such an audacious goal. I've been beyond lucky to have her as not only a source of confidence and encouragement, but also an endless wellspring of sharp sociological insight.

I can't imagine trying to write a book without all these people in my corner. I'm very happy that I didn't have to.

⨯ APPENDIX A ⨯
REGRESSION RESULTS TABLE

Logistic regression for Canadian attitudes on immigration

	"Overall, there is too much immigration to Canada."	"Many people claiming to be refugees are not actual refugees."	"On balance, immigrants to Canada are making Canada a worse place."
KEY MEASURE			
Alberta	1.393**	1.448**	2.086***
	(1.014, 1.904)	(1.066, 1.964)	(1.436, 2.989)
SEX			
Male	1.124	1.729***	1.389**
	(0.921, 1.371)	(1.431, 2.092)	(1.07, 1.807)
AGE (REFERENCE CATEGORY: 18 TO 29)			
30 to 39	1.063	0.867	1.099
	(0.662, 1.712)	(0.543, 1.385)	(0.531, 2.304)
40 to 49	1.651**	1.33	2.685***
	(1.065, 2.586)	(0.862, 2.066)	(1.463, 5.187)
50 to 59	1.394	1.947***	2.224**
	(0.905, 2.169)	(1.286, 2.978)	(1.212, 4.293)
60 to 69	1.427**	2.25***	1.807**
	(0.951, 2.166)	(1.522, 3.367)	(1.01, 3.42)
70 to 79	1.177	2.107***	1.712**
	(0.771, 1.815)	(1.408, 3.188)	(0.937, 3.293)

80 and over	1.001 (0.603, 1.662)	2.104*** (1.314, 3.389)	1.853** (0.926, 3.811)
Declined to state	1.86** (1.068, 3.242)	2.167*** (1.27, 3.712)	2.433** (1.143, 5.248)

EDUCATION (REFERENCE CATEGORY: COMPLETED UNIVERSITY)

No/some secondary school	3.042*** (2.084, 4.443)	1.774*** (1.224, 2.569)	2.438*** (1.492, 3.944)
Completed secondary school	2.578*** (1.907, 3.492)	1.928*** (1.44, 2.584)	2.136*** (1.428, 3.197)
Some college	1.553** (0.99, 2.4)	1.67** (1.099, 2.527)	1.843** (1.021, 3.21)
Completed college	1.793*** (1.343, 2.395)	1.983*** (1.508, 2.611)	1.833*** (1.245, 2.705)
Some university	1.164 (0.725, 1.828)	1.199 (0.78, 1.824)	1.889** (1.058, 3.263)
Postgraduate	0.864 (0.6, 1.232)	1.025 (0.742, 1.412)	0.747 (0.432, 1.247)
Declined to state	0.605 (0.193, 1.572)	0.877 (0.357, 2.009)	1.348 (0.368, 3.906)

POPULATION DENSITY (PERSONS PER KM^2; REFERENCE CATEGORY: 4,000 OR MORE)

Less than 500	0.876 (0.655, 1.171)	0.834 (0.635, 1.095)	0.968 (0.665, 1.411)
500 to 1,999	1.12 (0.838, 1.499)	0.938 (0.71, 1.237)	1.134 (0.782, 1.65)
2,000 to 3,999	1.003 (0.756, 1.33)	0.973 (0.747, 1.268)	0.794 (0.541, 1.167)

Note: Values odds ratios. Parenthetical values represent 95% confidence intervals. This analysis is based on data from the Environics Institute for Survey Research Focus Canada 2021 Survey. All computations were prepared by the author, and the responsibility for the use and interpretation of these data is entirely that of the author.

** $p < .01$
*** $p < .001$

✗ APPENDIX B ✗
NATURAL LANGUAGE PROCESSING

I came to computational methods as an outsider. When I began grad school, I dreaded my PhD program's required courses in quantitative methods, and once they were over, I did not give the tools we learned in them much further consideration as I continued my training in qualitative methods like interviews, ethnography, and content analysis. What held the door ajar for me until I circled back to quantitative and computational methods was mostly the work of the late John Mohr. I had the good fortune to take one of the final sociological theory seminars John taught before he passed away in 2019, and I left it not only with excitement about the study of culture but also with admiration for his openness to diverse methodological approaches. Though I did not imagine myself pursuing the methods at the time, I was inspired by the way that Mohr and his coauthors were able to marry computational tools with an inductive analytical approach that was familiar to me as a qualitative researcher.

When the world shut down in early 2020 due to the COVID-19 pandemic, I, like many other scholars, had difficulty imagining when the path to resuming the kind of research I was used to conducting would open. Institutional review boards had suspended approval for in-person research, and that ruled out many of the research activities I had planned. It was in this context that I made what Mohr and his coauthors of *Measuring Culture* refer to as a "research pivot" (Mohr et al. 2020) that involved taking up tools like computational text analysis to add breadth and depth to the qualitative analysis in this project. I was pleased to find that, despite being based on numbers, these approaches supported a rich, inductive, interpretive analysis not unlike what I was used to applying to qualitative data from interviews and participant observation. One aspect that makes computational methods, including the word embeddings approach utilized in chapter 5 of this book, a good supplement for qualitative methods of cultural analysis is that they are aligned with theoretical tenets such as the relationality (Kozlowski, Taddy, and Evans 2019; Stoltz and Taylor 2021) and multiplicity (Nelson 2021) of meaning systems.

Below is a detailed description of the analysis of *Los Angeles Times* articles on homelessness discussed in chapter 5. This approach uses tools from Natural Language Processing and other branches of data science to generate a holistic picture of the newspaper's coverage of homelessness. The resulting visualization provides insights into not only the patterns that characterize themes in that coverage, but also the overall structure of those themes and their relationships to each other.

To analyze the corpus of homelessness-related articles, I used word embeddings. Word embeddings are geometric representations of text that take a large corpus of data and output a model where each term is assigned a set of coordinates in a high-dimensional space. These coordinates are based on words' frequency of co-occurrence within a given researcher-defined "context window." The result is a model where the similarity of words' coordinates corresponds with their similarity or relatedness in the text data. Word embeddings can be used to solve analogy puzzles like "man is to king as woman is to what?" using arithmetic (subtracting the coordinates for "man" from those of "king" and then adding the coordinates for "woman" should yield coordinates similar to those for "queen").

I trained my word embeddings model on a corpus of *LA Times* articles that includes the homelessness-specific sample. I took this approach because word embeddings require a large quantity of text to produce robust results. This larger corpus ($n = 99,006$) features articles published in the paper within the same time frame (2015–2019). To maintain as consistent a semantic environment as possible, I selected only articles from news-related sections of the paper (e.g., "Politics," "World & Nation," and "California"), excluding parts of the paper like "Entertainment" and "Sports" that tend to be further afield from the articles in the homelessness sample. The text of the articles was captured using web scraping techniques built around the {rvest} (Wickham 2022) package for the R statistical programming language.

After cleaning the text by removing "stop words," I trained the resulting word embeddings model using the "global vectors" (GloVe) algorithm, as implemented in the R package {text2vec} (Selivanov, Bickel, and Wang 2022), with a symmetrical context window of 5. For the subsequent dimensionality reduction and clustering analysis, I included a subset of words contained in the corpus of homeless-specific articles. This involved first tokenizing and part-of-speech tagging the text using the SpaCy parser (Honnibal and Johnson 2015). I then retained only nouns, verbs, adjectives, and adverbs to filter out more meaningfully ambiguous parts of speech like interjections, pronouns, and conjunctions. From this filtered text, I calculated each word's term frequency-inverse document frequency (TF-IDF). TF-IDF is a measure of a word's importance to a document that considers the number of times it appears in the document and the total number of documents in the dataset in which it occurs. Words that score highest are words that appear frequently in a given document but infrequently overall. I then used the top ten words by TF-IDF for each article in the subsequent analysis.

After subsetting the word embeddings model to include only the embeddings for these words, the next step in the process was reducing the dimensionality of those embeddings from 300 to 2 so that it can be visualized as a scatterplot. There are several different methods available for dimensionality reduction. I chose Uniform Manifold Approximation and Projection (UMAP) following Nanni and Fallin (2021), who similarly use UMAP to visualize word embeddings trained on abstracts from academic articles related to climate change. As they note, UMAP is better suited than some of the other available techniques, including t-SNE (Maaten and Hinton 2008), for the task of identifying clusters (McInnes, Healy, and Melville 2018). Non-linear dimensionality reduction techniques like UMAP produce low-dimensional representations of the data that aim to preserve the overall structure of the data. In other words, points that are close to each other in the high-dimensional data should remain close in the low-dimensional representation.

Again following Nanni and Fallin (2021), I use Gaussian Mixture Modeling (GMM) to identify clusters within the new two-dimensional projection of the data. The implementation of GMM in the R package {mclust} (Scrucca et al. 2016) does not require that the researcher predetermine the number of clusters, g. Instead, the function fits models for a range of possible values for g and chooses the optimal model based on the Bayesian information criterion. I considered models up to a maximum value of 20 for g, which resulted in a final model containing the nine clusters analyzed in chapter 5.

A final step prior to visualizing the data involved assigning each word in the model a score for its "centrality" to its given cluster. To do so, I constructed a similarity matrix for each cluster based on its words' pairwise cosine similarity in the original word embeddings model. From this matrix, I then computed each word's eigenvector centrality. The final visualization includes the words with a centrality score greater than or equal to their cluster's median. The top twenty most central words are labeled. The opacity and size of both points and labels correspond with the word's centrality score, where darker points/labels are more central terms.

Identifying the articles that best exemplify each cluster requires calculating a single point in the 300-dimension embedding space that corresponds with each article and likewise calculating a single point to stand in for each cluster. To calculate the "proxy vectors" for articles, I computed the average of the coordinates for each unique word in the article, weighted by its TF-IDF score for the article. Similarly, I computed the proxy vectors for clusters by averaging the words assigned to them, weighted by their centrality scores. This resulted in nine 300-dimensional vectors corresponding with each cluster and 1,304 300-dimensional vectors corresponding with each article. I then calculated a 9 x 1,304 similarity matrix containing the pairwise cosine similarities between each article and each cluster. These cosine similarity scores model a given article's engagement with a given topic. Similarly to other Natural Language Processing methods aimed at identifying latent themes in a corpus (e.g., topic modeling), this procedure calculates scores for each combination of article and cluster, as it assumes that each article is composed of some combination of the topical clusters identified in the analysis. Thus, an article is assumed to engage multiple different clusters to varying degrees.

REFERENCES

"An Act to amend The Sexual Sterilization Act, SA." 1942, c 48, https://canlii.ca/t/5418w, retrieved July 3, 2023.
"Agricultural Pests Act, RSA." 1942, c 76, https://www.canlii.ca/t/53wdl, retrieved July 3, 2023.
Angelo, Hillary. 2013. "Bird in Hand: How Experience Makes Nature." *Theory and Society* 42(4):351–68.
Angelo, Hillary. 2021. *How Green Became Good: Urbanized Nature and the Making of Cities and Citizens*. University of Chicago Press.
Aptekar, Sofya. 2015. "Visions of Public Space: Reproducing and Resisting Social Hierarchies in a Community Garden." *Sociological Forum* 30(1):209–27. doi: 10.1111/socf.12152.
Alberta Department of Agriculture. 1951. *Annual Report 1951*. Provincial Archives of Alberta, item 2557, box 60, acc. 70.144, page 38.
Alberta Department of Agriculture. 1954. *Annual Report 1954*. Provincial Archives of Alberta, item 2716, box 62, acc. 70.414, page 32.
Alberta Department of Agriculture. 1952. *Annual Report 1952*. Provincial Archives of Alberta, item 2605, box 62, acc. 70.144, page 29.
Alberta Department of Agriculture 1957. *Annual Report 1957*. Provincial Archives of Alberta, item 2908, box 72, acc. 70.414, page 50.
Alberta Department of Agriculture. 1959. *Annual Report 1959*. Provincial Archives of Alberta, item 3101, box 76, acc. 70.144, page 50.
Arluke, Arnold, and Clinton Sanders. 1996. *Regarding Animals*. Temple University Press.
Bargheer, Stefan. 2018. *Moral Entanglements: Conserving Birds in Britain and Germany*. University of Chicago Press.
Barnard, Alex V. 2023. *Conservatorship: Inside California's System of Coercion and Care for Mental Illness*. Columbia University Press.
Bartram, Robin. 2021. "Cracks in Broken Windows: How Objects Shape Professional Evaluation." *American Journal of Sociology* 126(4):759–94. doi: 10.1086/713763.
Beckett, Katherine, and Steve Herbert. 2009. *Banished: The New Social Control in Urban America*. Oxford University Press.

Bell, Michael. 2018. *City of the Good: Nature, Religion, and the Ancient Search for What Is Right.* Princeton University Press.

Bell, Shannon Elizabeth. 2013. *Our Roots Run Deep as Ironweed: Appalachian Women and the Fight for Environmental Justice.* University of Illinois Press.

Benjamin, Ruha. 2016a. "Informed Refusal: Toward a Justice-Based Bioethics." *Science, Technology and Human Values* 41(6):967–90. doi: 10.1177/0162243916656059.

Benjamin, Ruha. 2016b. "Interrogating Equity: A Disability Justice Approach to Genetic Engineering." *Issues in Science and Technology; Washington* 32(3):51–54.

Bennhold, Katrin. 2019. "A Fairy-Tale Baddie, the Wolf, Is Back in Germany, and Anti-Migrant Forces Pounce." *New York Times.* https://www.nytimes.com/2019/04/23/world/europe/germany-wolves-afd-immigration.html, retrieved June 13, 2019.

Biehler, Dawn. 2013. *Pests in the City: Flies, Bedbugs, Cockroaches, and Rats.* University of Washington Press.

Biehler, Dawn Day, and Gregory L. Simon. 2010. "The Great Indoors: Research Frontiers on Indoor Environments as Active Political-Ecological Spaces." *Progress in Human Geography* 35(2):172–92. doi: 10.1177/0309132510376851.

Billings, P. R., R. Hubbard, and S. A. Newman. 1999. "Human Germline Gene Modification: A Dissent." *Lancet* 353(9167):1873–75. doi: 10.1016/S0140-6736(99)01173-3.

Bloom, Tracy. 2019. "Mountain Lion P-47 Found Dead in Santa Monica Mountains; Rat Poison Suspected: NPS." *KTLA.* https://ktla.com/news/money-smart/mountain-lion-p-47-found-dead-in-santa-monica-mountains-after-ingesting-rat-poison-nps/, retrieved February 27, 2023.

Blue, Gwendolyn. 2008. "If It Ain't Alberta, It Ain't Beef: Local Food, Regional Identity, (Inter) National Politics." *Food, Culture & Society* 11(1):69–85. doi: 10.2752/155280108X276168.

Bourdieu, Pierre. 1986. *Distinction: A Social Critique of the Judgement of Taste.* Routledge & Kegan Paul.

Brouns, Stan J. J., Matthijs M. Jore, Magnus Lundgren, Edze R. Westra, Rik J. H. Slijkhuis, Ambrosius P. L. Snijders, Mark J. Dickman, Kira S. Makarova, Eugene V. Koonin, and John van der Oost. 2008. "Small CRISPR RNAs Guide Antiviral Defense in Prokaryotes." *Science* 321(5891):960–64. doi: 10.1126/science.1159689.

Brown, Paul. 2004. "Scientists Held Hostage on Darwin's Island." *The Guardian,* February 28.

Bullard, Robert D. 2018. *Dumping in Dixie: Race, Class, and Environmental Quality,* 3rd ed. Routledge.

Burawoy, Michael, Joseph A. Blum, Sheba George, Zsuzsa Gille, and Millie Thayer. 2000. *Global Ethnography: Forces, Connections, and Imaginations in a Postmodern World.* University of California Press.

Cairney, R. 1996. "'Democracy Was Never Intended for Degenerates': Alberta's Flirtation with Eugenics Comes Back to Haunt It." *CMAJ: Canadian Medical Association Journal* 155(6):789–92.

Cairns, Robert. 1992. "Natural Resources and Canadian Federalism: Decentralization, Recurring Conflict, and Resolution." *Publius: The Journal of Federalism* 22(1):55–70. doi: 10.1093/oxfordjournals.pubjof.a037996.

Campbell, Karl, and C. Josh Donlan. 2005. "Feral Goat Eradications on Islands." *Conservation Biology* 19(5):1362–74. doi: 10.1111/j.1523-1739.2005.00228.x.

Campbell, Karl J., Greg S. Baxter, Peter J. Murray, Bruce E. Coblentz, C. Josh Donlan, and Victor G. Carrion. 2005. "Increasing the Efficacy of Judas Goats by Sterilisation and Pregnancy Termination." *Wildlife Research* 32(8):737–43. doi: 10.1071/WR05033.

Canadian Census. 2006. "Population and Dwelling Counts, for Canada and Census Subdivisions, 2006 and 2001 Censuses." https://www12.statcan.gc.ca/census-recensement/2006/dp-pd/hlt/97-550/Index.cfm, retrieved January 1, 2018.

Canadian Census. 2016a. "Census Profile, 2016 Census Calgary, City [Census Subdivision], Alberta and Division No. 6, Census Division [Census Division], Alberta." https://www12.statcan.gc.ca/census-recensement/2016/dp-pd/prof/details/page.cfm, retrieved August 20, 2018.

Canadian Census. 2016b. "Percentage Visible Minorities for Canada, Provinces and Territories, 1996, 2001 and 2006 Censuses." https://www12.statcan.gc.ca/census-recensement/2006/dp-pd/92-596/P2-2.cfm, retrieved August 20, 2018.

Carney, Nikita. 2017. "Multi-Sited Ethnography: Opportunities for the Study of Race." *Sociology Compass* 11(9):e12505. doi: 10.1111/soc4.12505.

Catton, William R., and Riley Dunlap. 1980. "A New Ecological Paradigm for Post-Exuberant Sociology." *The American Behavioral Scientist* 24(1):15–47.

Catton, William R., and Riley E. Dunlap. 1978. "Environmental Sociology: A New Paradigm." *The American Sociologist* 13(1):41–49.

Census Bureau. 2021. "U.S. Census Bureau QuickFacts: Los Angeles City, California." https://www.census.gov/quickfacts/fact/table/losangelescitycalifornia/PST045221, retrieved September 5, 2022.

Chakrabarty, Dipesh. 2009. "The Climate of History: Four Theses." *Critical Inquiry* 35(2):197–222. doi: 10.1086/596640.

Charles Darwin Foundation and Galápagos National Park. 1997. "Final Report of the Galápagos Workshop: Feral Goat Eradication Program for Isla Isabela September 9–18, 1997."

Cherry, Elizabeth. 2019. *For the Birds: Protecting Wildlife through the Naturalist Gaze*. Rutgers University Press,

Chiland, Elijah. 2019. "LA Settles Homeless Property Rights Case." *Curbed LA*. https://la.curbed.com/2019/3/6/18253888/mitchell-settlement-los-angeles-homeless-property, retrieved July 28, 2023.

Cong, Le, F. Ann Ran, David Cox, Shuailiang Lin, Robert Barretto, Naomi Habib, Patrick D. Hsu, Xuebing Wu, Wenyan Jiang, Luciano A. Marraffini, and Feng Zhang. 2013. "Multiplex Genome Engineering Using CRISPR/Cas Systems." *Science* 339(6121):819–23. doi: 10.1126/science.1231143.

Cresswell, Tim. 2006. *On the Move: Mobility in the Modern Western World*. Taylor & Francis.

Cresswell, Tim. 2014. *Place: An Introduction*. Wiley.

Cribbs, Adam P., and Sumeth M. W. Perera. 2017. "Science and Bioethics of CRISPR-Cas9 Gene Editing: An Analysis Towards Separating Facts and Fiction." *Yale Journal of Biology and Medicine* 90(4):625–34.

Cronon, William. 1996. "The Trouble with Wilderness or Getting Back to the Wrong Nature." In *Uncommon Ground: Rethinking the Human Place in Nature*, pp. 69–90. W. W. Norton and Company.

Crutzen, Paul, and Eugene Stoermer. 2000. "The 'Anthropocene.'" *International Geosphere-Biosphere Programme Newsletter* 41:17–18.

Davis, Julie Hirschfeld. 2018. "Trump Calls Some Unauthorized Immigrants 'Animals' in Rant." *New York Times*. https://www.nytimes.com/2018/05/16/us/politics/trump-undocumented-immigrants-animals.html, retrieved February 15, 2019.

Davis, Mike. 1998. *Ecology of Fear: Los Angeles and the Imagination of Disaster*. Macmillan.

Do, Anh, Cindy Carcamo, and Joseph Serna. 2018. "O.C. Pushes Homeless off the Street but Can't Find Anywhere to Shelter Them." *Los Angeles Times*. https://www.latimes.com/local/lanow/la-me-oc-homeless-test-20180328-story.html, retrieved July 10, 2023.

Domínguez Rubio, Fernando. 2014. "Preserving the Unpreservable: Docile and Unruly Objects at MoMA." *Theory and Society* 43(6):617–45.

Domínguez Rubio, Fernando. 2016. "On the Discrepancy between Objects and Things: An Ecological Approach." *Journal of Material Culture* 21(1):59–86. doi: 10.1177/1359183515624128.

Domínguez Rubio, Fernando. 2020. *Still Life: Ecologies of the Modern Imagination at the Art Museum*. University of Chicago Press.

Douglas, M. 1966. *Purity and Danger: An Analysis of Concept of Pollution and Taboo*. Routledge.

Douglas, Thomas, and Julian Savulescu. 2010. "Synthetic Biology and the Ethics of Knowledge." *Journal of Medical Ethics* 36(11):687–93. doi: 10.1136/jme.2010.038232.

DTLA Alliance for Human Rights. n.d. "Who We Are." *LA Alliance for Human Rights*. https://www.la-alliance.org/who_we_are, retrieved July 28, 2023.

Dunlap, Riley, and William Catton. 1983. "What Environmental Sociologists Have in Common (whether Concerned with "Built" or "Natural" Environments)." *Sociological Inquiry* 553(2–3):113–35.

Durando, Jessica. 2019. "Donald Trump Jr. Compares Border Wall to Zoo in an Instagram Post." *USA Today*. https://www.usatoday.com/story/news/politics/2019/01/09/donald-trump-jr-compares-border-wall-zoo-instagram-post/2523385002, retrieved March 15, 2019.

Durkheim, Émile. 2014. *The Rules of Sociological Method: And Selected Texts on Sociology and Its Method*. Simon and Schuster.

Elliott, Rebecca. 2018. "The Sociology of Climate Change as a Sociology of Loss." *European Journal of Sociology / Archives Européennes de Sociologie* 59(3):301–37. doi: 10.1017/S0003975618000152.

Esvelt, Kevin M., Andrea L. Smidler, Flaminia Catteruccia, and George M. Church. 2014. "Concerning RNA-Guided Gene Drives for the Alteration of Wild Populations." *Elife* 3. doi: 10.7554/eLife.03401.

Farrell, Justin. 2015. *The Battle for Yellowstone: Morality and the Sacred Roots of Environmental Conflict*. Princeton University Press.

Finnegan, Michael. 2019. "Julian Castro Calls for Surge in Federal Spending to End Homelessness." *Los Angeles Times*. https://www.latimes.com/politics/la-na-pol-2020-castro-housing-homelessness-presidential-20190617-story.html, retrieved July 10, 2023.

Freudenburg, William R., Scott Frickel, and Robert Gramling. 1995. "Beyond the Nature/Society Divide: Learning to Think about a Mountain." *Sociological Forum* 10(3):361–92.

Freudenburg, William R., and Robert Gramling. 1993. "Socioenvironmental Factors and Development Policy: Understanding Opposition and Support for Offshore Oil." *Sociological Forum* 8(3):341.

Galápagos Conservancy. n.d. "Project Isabela." *Galápagos Conservancy*. https://www.galapagos.org/conservation/project-isabela/, retrieved June 5, 2023.

Gammon, Katharine. 2022. "'He Changed Us': The Remarkable Life of Celebrity Mountain Lion P-22." *The Guardian*, December 20.

Gantz, Valentino M., and Ethan Bier. 2015. "Genome Editing. The Mutagenic Chain Reaction: A Method for Converting Heterozygous to Homozygous Mutations." *Science* 348(6233):442–44. doi: 10.1126/science.aaa5945.

Ghaziani, Amin. 2009. "An 'Amorphous Mist'? The Problem of Measurement in the Study of Culture." *Theory and Society* 38(6):581–612.

Gibbins, Roger. 1980. *Prairie Politics and Society: Regionalism in Decline*. Butterworths.

Gibbins, Roger. 1992. *Alberta and the National Community*. University of Alberta Press.

Gong, Neil. 2021. "Perspective | California Gave People the 'Right' to Be Homeless—but Little Help Finding Homes." *Washington Post*, May 20.

Gorman, Anna, and Harriet Blair Rowan. 2019. "L.A. County's Homeless Population Is Growing—but Not as Fast as They're Dying." *Los Angeles Times*. https://www.latimes.com/local/california/la-me-ln-homeless-people-death-unsheltered-substance-abuse-20190422-story.html, retrieved July 9, 2023.

Gray, Summer. 2023. *In the Shadow of the Seawall: Coastal Injustice and the Dilemma of Placekeeping*. University of California Press.

Grazian, David. 2015. *American Zoo: A Sociological Safari*. Princeton University Press.

Grekul, Jana. 2008. "Sterilization in Alberta, 1928 to 1972: Gender Matters." *Canadian Review of Sociology/Revue Canadienne de Sociologie* 45(3):247–66. doi: 10.1111/j.1755-618X.2008.00014.x.

Grekul, Jana, Arvey Krahn, and Dave Odynak. 2004. "Sterilizing the 'Feeble-Minded': Eugenics in Alberta, Canada, 1929–1972." *Journal of Historical Sociology* 17(4):358–84. doi: 10.1111/j.1467-6443.2004.00237.x.

Griswold, Wendy. 1986. *Renaissance Revivals: City Comedy and Revenge Tragedy in the London Theater, 1576–1980*. University of Chicago Press.

Harvey, David. 1996. *Justice, Nature and the Geography of Difference*. Wiley.

Hassig, Debra. 1995. *Medieval Bestiaries: Text, Image, Ideology*. Cambridge University Press.

Herring, Chris. 2014. "The New Logics of Homeless Seclusion: Homeless Encampments in America's West Coast Cities." *City & Community* 13(4):285–309. doi: 10.1111/cico.12086.

Herring, Chris. 2019. "Complaint-Oriented Policing: Regulating Homelessness in Public Space." *American Sociological Review* 84(5):769–800. doi: 10.1177/0003122419872671.

Heynen, Nikolas C., Maria Kaika, and Erik Swyngedouw, eds. 2006. *In the Nature of Cities: Urban Political Ecology and the Politics of Urban Metabolism*. Taylor & Francis.

Hilbert, Richard A. 1994. "People Are Animals: Comment on Sanders and Arluke's 'If Lions Could Speak.'" *Sociological Quarterly* 35(3):533–36. doi: 10.1111/j.1533-8525.1994.tb01744.x.

"History of Rat Control in Alberta." n.d. https://www.alberta.ca/history-of-rat-control-in-alberta.aspx, retrieved May 29, 2023.

Holland, Gale. 2015a. "How the Homeless Live and What They Keep in L.A." *Los Angeles Times*. https://www.latimes.com/local/california/la-me-homeless-belongings-20150616-story.html, retrieved July 10, 2023.

Holland, Gale. 2015b. "More Homeless Camps Are Appearing beyond Downtown

L.A.'s Skid Row." *Los Angeles Times*. https://www.latimes.com/local/california/la-me-homeless-encampments-20150125-story.html, retrieved July 10, 2023.

Holland, Gale. 2017. "U.N. Monitor Says L.A. Lags behind Other Cities in Attacking Homelessness." *Los Angeles Times*. https://www.latimes.com/local/lanow/la-me-ln-un-monitor-skid-row-20171215-story.html, retrieved July 10, 2023.

Honnibal, Matthew, and Mark Johnson. 2015. "An Improved Non-Monotonic Transition System for Dependency Parsing." In *Proceedings of the 2015 Conference on Empirical Methods in Natural Language Processing*, pp. 1373–78. Association for Computational Linguistics.

Hoyman, Michele M., and Jamie R. McCall. 2013. "The Evolution of Ecotourism: The Story of the Galapagos Islands and the Special Law of 1998." In *Science and Conservation in the Galapagos Islands: Frameworks & Perspectives, Social and Ecological Interactions in the Galapagos Islands*, ed. S. J. Walsh and C. F. Mena, pp. 127–40. Springer.

Irvine, Leslie. 2008. *If You Tame Me: Understanding Our Connection with Animals*. Temple University Press.

Ishino, Y., H. Shinagawa, K. Makino, M. Amemura, and A. Nakata. 1987. "Nucleotide Sequence of the Iap Gene, Responsible for Alkaline Phosphatase Isozyme Conversion in Escherichia Coli, and Identification of the Gene Product." *Journal of Bacteriology* 169(12):5429–33.

"Jap Trap." 1941–1945. National Archives and Records Administration Collection. ddr-densho-37-498. Densho Digital Archives. https://ddr.densho.org/ddr-densho-37-498/.

Jerolmack, Colin. 2008. "How Pigeons Became Rats: The Cultural-Spatial Logic of Problem Animals." *Social Problems* 55(1):72–94. doi: 10.1525/sp.2008.55.1.72.

Jerolmack, Colin. 2013. *The Global Pigeon*. University of Chicago Press.

Jerolmack, Colin, and Iddo Tavory. 2014. "Molds and Totems: Nonhumans and the Constitution of the Social Self." *Sociological Theory* 32(1):64–77. doi: 10.1177/0735275114523604.

Karlamangla, Soumya. 2019. "Long before City Hall Rats, L.A. Has Struggled with the Rise of Typhus." *Los Angeles Times*. https://www.latimes.com/local/california/la-me-ln-typhus-20190217-story.html, retrieved February 27, 2023.

Kozlowski, Austin C., Matt Taddy, and James A. Evans. 2019. "The Geometry of Culture: Analyzing the Meanings of Class through Word Embeddings." *American Sociological Review* 84(5):905–49. doi: 10.1177/0003122419877135.

Krause, William J., and Winifred A. Krause. 2006. *The Opossum: Its Amazing Story*. University of Missouri-Columbia, Department of Pathology and Anatomical Sciences, School of Medicine.

Kubal, Timothy J. 1998. "The Presentation of Political Self: Cultural Resonance and the Construction of Collective Action Frames." *Sociological Quarterly* 39(4):539–54. doi: 10.1111/j.1533-8525.1998.tb00517.x.

Kuhn, Thomas S. 1962. *The Structure of Scientific Revolutions*. University of Chicago Press.

Kuykendall, Kate. 2014. "Griffith Park Mountain Lion Exposed to Poison, Suffering from Mange—Santa Monica Mountains National Recreation Area (U.S. National Park Service)." National Park Service. https://www.nps.gov/samo/learn/news/gp-lion-exposed-to-poison.htm, retrieved February 27, 2023.

Kuykendall, Kate. 2015. "Mountain Lion Was Exposed to Multiple Poisons, Tests Show—Santa Monica Mountains National Recreation Area (U.S. National Park Service)." *National Park Service*. https://www.nps.gov/samo/learn/news/p-34-test-results-confirmed.htm, retrieved February 27, 2023.

Lamont, Michèle. 1995. "National Identity and National Boundary Patterns in France and the United States." *French Historical Studies* 19(2):349–65. doi: 10.2307/286776.

Latour, Bruno. 1988. *The Pasteurization of France*. Harvard University Press.

Law, John. 1987. "On the Social Explanation of Technical Change: The Case of the Portuguese Maritime Expansion." *Technology and Culture* 28(2):227–52. doi: 10.2307/3105566.

Lewis, Simon L., and Mark A. Maslin. 2015. "Defining the Anthropocene." *Nature* 519(7542):171–80. doi: 10.1038/nature14258.

Lock, Margaret. 2013a. "The Epigenome and Nature/Nurture Reunification: A Challenge for Anthropology." *Medical Anthropology* 32(4):291–308. doi: 10.1080/01459740.2012.746973.

Lock, Margaret. 2013b. "The Lure of the Epigenome." *Lancet* 381(9881):1896–97. doi: 10.1016/S0140-6736(13)61149-6.

Lock, Margaret. 2015. "Comprehending the Body in the Era of the Epigenome." *Current Anthropology* 56(2):151–77. doi: 10.1086/680350.

Lock, Margaret. 2017. "Recovering the Body." *Annual Review of Anthropology* 46(1):1–14. doi: 10.1146/annurev-anthro-102116-041253.

Lopez, Steve. 2017. "Column: What If This Homeless Woman Were Your Mother—Would You Keep Moving or Step in to Help?" *Los Angeles Times*. https://www.latimes.com/local/california/la-me-lopez-gravely-disabled-20171101-story.html, retrieved July 8, 2023.

Lopez, Steve. 2019a. "Column: As L.A.'s Homeless Crisis Worsens, No One Is in Charge. That Has to Change." *Los Angeles Times*. https://www.latimes.com/california/story/2019-11-20/a-call-for-new-strategies-and-solutions-on-homelessness, retrieved July 10, 2023.

Lopez, Steve. 2019b. "Column: The City Has a Growing Mountain of Possessions Confiscated from Homeless People." *Los Angeles Times*. https://www.latimes.com/california/story/2019-07-30/homeless-possessions-storage-skid-row-steve-lopez-column, retrieved July 10, 2023.

Los Angeles Homeless Services Authority. 2020. "Greater Los Angeles Homeless Count—City of Los Angeles." https://www.lahsa.org/documents?id=4680-2020-greater-los-angeles-homeless-count-city-of-los-angeles, retrieved September 5, 2022.

Los Angeles Times. 1994. "Trackers Hunt Mountain Lion That Killed Jogger." *Los Angeles Times*. https://www.latimes.com/archives/la-xpm-1994-04-27-mn-50967-story.html, retrieved July 5, 2023.

Loughran, Kevin. 2017. "Race and the Construction of City and Nature." *Environment and Planning A: Economy and Space* 49(9):1948–67. doi: 10.1177/0308518X17713995.

Lyytimäki, Jari. 2014. "Bad Nature: Newspaper Representations of Ecosystem Disservices." *Urban Forestry & Urban Greening* 13(3):418–24. doi: 10.1016/j.ufug.2014.04.005.

Maaten, Laurens van der, and Geoffrey Hinton. 2008. "Visualizing Data Using T-SNE." *Journal of Machine Learning Research* 9(86):2579–605.

Mali, Prashant, Luhan Yang, Kevin M. Esvelt, John Aach, Marc Guell, James E. DiCarlo, Julie E. Norville, and George M. Church. 2013. "RNA-Guided Human Genome Engineering via Cas9." *Science* 339(6121):823–26. doi: 10.1126/science.1232033.

Manning, Ernest. 1953–1955a. "Addresses and Speeches." Premier Papers, Provincial Archives of Alberta, PR1969.0289/1826. 40.72.

Manning, Ernest. 1953–1955b. "Articles." Premier Papers, Provincial Archives of Alberta, PR1969.0289/1920. 86.8.

Martinez, Christian. 2022. "P-22, L.A.'s Celebrity Mountain Lion, Captured in the Backyard of a Los Feliz Home." *Los Angeles Times*. https://www.latimes.com/california/story/2022-12-12/p-22-captured-in-backyard-of-los-feliz-home-resident-says, retrieved February 27, 2023.

Massey, Doreen. 1994. *Space, Place, and Gender*. University of Minnesota Press.

Mather, Kate. 2016a. "L.A. Police Chief Beck Backs Charges against Officer Who Fatally Shot Venice Homeless Man." *Los Angeles Times*. https://www.latimes.com/local/crime/la-me-venice-shooting-20160112-story.html, retrieved July 10, 2023.

Mather, Kate. 2016b. "LAPD Chief Recommends Criminal Charges for Officer in Fatal Shooting of Homeless Man in Venice." *Los Angeles Times*. https://www.latimes.com/local/lanow/la-me-ln-lapd-shooting-venice-brendan-glenn-20160111-story.html, retrieved July 10, 2023.

Mather, Kate, Cindy Chang, and Marisa Gerber. 2018. "D.A. Declines to Charge Former LAPD Officer in Fatal Shooting of Homeless Man near Venice Boardwalk." *Los Angeles Times*. https://www.latimes.com/local/lanow/la-me-ln-lapd-shooting-glenn-20180307-story.html, retrieved July 10, 2023.

Mayorga-Gallo, Sarah. 2018. "Whose Best Friend? Dogs and Racial Boundary Maintenance in a Multiracial Neighborhood." *Sociological Forum* 33(2):505–28. doi: 10.1111/socf.12425.

McCumber, Andrew. 2018. "Producing the Natural: Mobilities and the Managed Aesthetics of Nature." *Sociology Compass* 12(9):e12619. doi: 10.1111/soc4.12619.

McCumber, Andrew, and Patrick Neil Dryden. 2022. "The Bestiary in the Candy Aisle: A Framework for Nature in Unexpected Places." *Environmental Humanities* 14(1):110–28. doi: 10.1215/22011919-9481462.

McDonald, Jeff. 2017. "Homeless Man Wins Harassment Settlement from San Diego Police." *Los Angeles Times*. https://www.latimes.com/local/lanow/la-me-sd-police-harassment-20170214-story.html, retrieved July 10, 2023.

McDonnell, Terence E. 2010. "Cultural Objects as Objects: Materiality, Urban Space, and the Interpretation of AIDS Campaigns in Accra, Ghana." *American Journal of Sociology* 115(6):1800–1852. doi: 10.1086/651577.

McDonnell, Terence E. 2014. "Drawing out Culture: Productive Methods to Measure Cognition and Resonance." *Theory and Society* 43(3/4):247–74.

McDonnell, Terence E. 2016. *Best Laid Plans: Cultural Entropy and the Unraveling of AIDS Media Campaigns*. University of Chicago Press.

McDonnell, Terence E. 2023. "Cultural Objects, Material Culture, and Materiality." *Annual Review of Sociology* 49(1): 195–220. doi: 10.1146/annurev-soc-031021-041439.

McDonnell, Terence E., Christopher A. Bail, and Iddo Tavory. 2017. "A Theory of Resonance." *Sociological Theory* 35(1):1–14. doi: 10.1177/0735275117692837.

McInnes, Leland, John Healy, Nathaniel Saul, and Lucas Großberger. 2018. "UMAP: Uniform Manifold Approximation and Projection for Dimension Reduction." *Journal of Open Source Software* 3(29):861. doi: 10.21105/joss.00861.

McLaren, Angus. 1990. *Our Own Master Race: Eugenics in Canada, 1885–1945*. McClelland & Stewart.
McMahan, Jeff. 2010. "The Meat Eaters." *New York Times*, September 10.
McTavish, Lianne, and Jingjing Zheng. 2011. "Rats in Alberta: Looking at Pest-Control Posters from the 1950s." *Canadian Historical Review* 92(3):515–46. doi: 10.1353/can.2011.0045.
Mills, C. Wright. 2000. *The Sociological Imagination*. Oxford University Press.
Mitchell v. City of Los Angeles. 2016. Case no. 2:16-cv-01750 (C.D. Cal. 2016).
Mohai, Paul, David Pellow, and J. Timmons Roberts. 2009. "Environmental Justice." *Annual Review of Environment & Resources* 34(1):405–30. doi: 10.1146/annurev-environ-082508-094348.
Mohr, John W., Christopher A. Bail, Margaret Frye, Jennifer C. Lena, Omar Lizardo, Terence E. McDonnell, Ann Mische, Iddo Tavory, and Frederick F. Wherry. 2020. *Measuring Culture*. Columbia University Press.
Money, Luke. 2018. "Orange County Leaders Want to Use Mental Health Facility in Costa Mesa to Shelter Homeless." *Los Angeles Times*. https://www.latimes.com/local/lanow/la-me-oc-homeless-20180324-story.html, retrieved July 10, 2023.
Mulvihill, John J., Benjamin Capps, Yann Joly, Tamra Lysaght, Hub A. E. Zwart, Ruth Chadwick, and Law International Human Genome Organisation (HUGO) Committee of Ethics and Society (CELS). 2017. "Ethical Issues of CRISPR Technology and Gene Editing through the Lens of Solidarity." *British Medical Bulletin* 122(1):17–29. doi: 10.1093/bmb/ldx002.
Murphy, Raymond, and Riley Dunlap. "Beyond the Society/Nature Divide: Building on the Sociology of William Freudenburg." *Journal of Environmental Studies and Sciences* 2(1):7–17.
Nading, Alex M. 2014. *Mosquito Trails: Ecology, Health, and the Politics of Entanglement*. University of California Press.
Nanni, Antonio, and Mallory Fallin. 2021. "Earth, Wind, (Water), and Fire: Measuring Epistemic Boundaries in Climate Change Research." *Poetics* 88:101573. doi: 10.1016/j.poetic.2021.101573.
National Academies of Sciences. 2016. *Gene Drives on the Horizon: Advancing Science, Navigating Uncertainty, and Aligning Research with Public Values*.
Nelson, Laura K. 2021. "Leveraging the Alignment between Machine Learning and Intersectionality: Using Word Embeddings to Measure Intersectional Experiences of the Nineteenth Century U.S. South." *Poetics* 88:101539. doi: 10.1016/j.poetic.2021.101539.
Nicholls, Henry. 2014. *The Galapagos: A Natural History*. Basic Books.
Nixon, Rob. 2011. *Slow Violence and the Environmentalism of the Poor*. Harvard University Press.
Omi, Michael, and Howard Winant. 2014. *Racial Formation in the United States*. Routledge.
Pachirat, Timothy. 2011. *Every Twelve Seconds: Industrialized Slaughter and the Politics of Sight*. Yale University Press.
Partovi, Susan. 2019. "Op-Ed: My Patient Was Homeless. I Knew She Was Going to Die, but My Hands Were Tied." *Los Angeles Times*. https://www.latimes.com/opinion/op-ed/la-oe-partovi-homeless-deaths-treatment-20190505-story.html, retrieved July 8, 2023.
Pellow, David N., and Hollie Nyseth Brehm. 2013. "An Environmental Sociology for

the Twenty-First Century." *Annual Review of Sociology* 39:229–50. doi: 10.1146/annurev-soc-071312-145558.

Pellow, David Naguib. 2017. *What Is Critical Environmental Justice?* Wiley.

Penfold, Steven. 2002. "Eddie Shack Was No Tim Horton: Donuts and the Folklore of Mass Culture in Canada." In *Food Nations: Selling Taste in Consumer Society*, ed. Warren Belasco and Philip Scranton, pp. 48–66. Routledge.

Plumer, Brad, Javier Zarracina, Eliza Barclay, Julia Belluz, and Umair Irfan. 2016. "A Simple Guide to CRISPR, One of the Biggest Science Stories of the Decade." https://www.vox.com/science-and-health/2016/12/30/13164064/crispr-cas9-gene-editing, retrieved April 27, 2018.

Price, Jenny. 2006. "Thirteen Ways of Seeing Nature in LA." *The Believer*.

Queally, James, and Laura J. Nelson. 2022. "'Incredibly Difficult': Why Officials Euthanized Ailing Mountain Lion P-22." *Los Angeles Times*. https://www.latimes.com/california/story/2022-12-17/why-officials-euthanized-ailing-mountain-lion-p22, retrieved February 27, 2023.

Reyes-Velarde, Alejandra. 2018. "L.A. River Bike Path Has Become Lure for Homeless, Reseda Residents Say." *Los Angeles Times*. https://www.latimes.com/local/lanow/la-me-reseda-bike-path-20171204-story.html, retrieved July 10, 2023.

Robnett, Belinda. 2004. "Emotional Resonance, Social Location, and Strategic Framing." *Sociological Focus* 37(3):195–212.

Rodriguez, E. 2016. "Ethical Issues in Genome Editing Using Crispr/Cas9 System." *Journal of Clinical Research & Bioethics* 7(2):1–4. doi: 10.4172/2155-9627.1000266.

Russell, Edmund. 2001. *War and Nature: Fighting Humans and Insects with Chemicals from World War I to Silent Spring*. Cambridge University Press.

Sanders, C. 2010. *Understanding Dogs*. Temple University Press.

Sanders, Clinton, and Arnold Arluke. 1993. "If Lions Could Speak: Investigating the Animal-Human Relationship and the Perspectives of Nonhuman Others." *Sociological Quarterly* 34(3):377–90.

Schemo, Diana Jean. 1995. "Galapagos Islands Journal; Homo Sapiens at War on Darwin's Peaceful Isles." *New York Times*, November 28.

Schwarz, Ori. 2015. "The Sound of Stigmatization: Sonic Habitus, Sonic Styles, and Boundary Work in an Urban Slum." *American Journal of Sociology* 121(1):205–42. doi: 10.1086/682023.

Scoville, Caleb. 2022. "Constructing Environmental Compliance: Law, Science, and Endangered Species Conservation in California's Delta." *American Journal of Sociology* 127(4):1094–1150. doi: 10.1086/718277.

Scrucca, Luca, Michael Fop, T. Brendan Murphy, and Adrian E. Raftery. 2016. "Mclust 5: Clustering, Classification and Density Estimation Using Gaussian Finite Mixture Models." *The R Journal* 8(1):289–317.

Selivanov, Dmitriy, Manuel Bickel, and Qing Wang. 2022. "Text2vec: Modern Text Mining Framework for R." http://text2vec.org, retrieved July 11, 2023.

Sennett, Richard. 1970. *The Uses of Disorder: Personal Identity & City Life*. Knopf.

Serna, Joseph, and Hailey Branson-Potts. 2016. "Griffith Park Mountain Lion P-22 Suspected of Killing Koala at L.A. Zoo." *Los Angeles Times*. https://www.nps.gov/samo/learn/news/gp-lion-exposed-to-poison.htm, retrieved February 27, 2023.

Sewell, Abby. 2016. "Los Angeles County Outlines Strategies to Reduce Homelessness."

Los Angeles Times. https://www.latimes.com/local/lanow/la-me-ln-homeless-plans-20160107-story.html, retrieved July 10, 2023.

Sewell, Alyasah Ali. 2020. "Policing the Block: Pandemics, Systemic Racism, and the Blood of America." *City & Community* 19(3):496–505. doi: 10.1111/cico.12517.

"The Sexual Sterilization Act." SA March 21, 1928, c 37, https://canlii.ca/t/53zws, retrieved on July 3, 2023.

Sheller, Mimi. 2016. "Mobility, Freedom and Public Space." In *The Ethics of Mobilities: Rethinking Place, Exclusion, Freedom and Environment*, ed. S. Bergmann and T. Sager, pp. 41–54. London: Routledge.

Sheller, Mimi, and John Urry. 2006. "The New Mobilities Paradigm." *Environment and Planning A* 38(2):207–26. doi: 10.1068/a37268.

Smith, Doug. 2017. "The Biggest Barrier to Opening a Homeless Shelter in L.A.? Location, Location, Location." *Los Angeles Times.* https://www.latimes.com/local/lanow/la-me-ln-shelter-challenges-20170904-story.html, retrieved July 10, 2023.

Smith, Dakota. 2019. "Garcetti's Budget Would Spend More Money on Street Repairs and Homelessness." *Los Angeles Times.* https://www.latimes.com/local/lanow/la-me-ln-budget-2019-eric-garcetti-20190418-story.html, retrieved July 10, 2023.

Smith, Dakota, and David Zahniser. 2019 "Filth from Homeless Camps Is Luring Rats to L.A. City Hall, Report Says." *Los Angeles Times.* https://www.latimes.com/local/lanow/la-me-ln-rats-homelessness-city-hall-fleas-report-20190603-story.html, retrieved June 13, 2023.

Smith, Doug. 2016a. "A Fix for L.A.'s Homeless Crisis Isn't Cheap. Will Voters Go for $1.2 Billion in Borrowing?" *Los Angeles Times.* https://www.latimes.com/local/lanow/la-me-ln-homeless-tax-measure-20160913-snap-story.html, retrieved July 10, 2023.

Smith, Doug. 2016b. "With El Niño Danger Passed, Focus Shifts on Homeless River Dwellers." *Los Angeles Times.* https://www.latimes.com/local/california/la-me-river-dwellers-20160424-story.html, retrieved July 10, 2023.

Smith, Joshua Emerson. 2017. "San Diego Focusing on Homeless Camps along River Following Hepatitis A Outbreak." *Los Angeles Times.* https://www.latimes.com/local/lanow/la-me-ln-san-diego-hepatitis-outbreak-20171105-story.html, retrieved July 10, 2023.

Solis, Nathan. 2022a. "Death of Pregnant Mountain Lion Underscores Two Human-Caused Dangers: Cars and Rat Poison." *Los Angeles Times.* https://www.latimes.com/california/story/2022-09-07/mountain-lion-p-54-pregnant-when-struck-killed-by-vehicle, retrieved February 27, 2023.

Solis, Nathan. 2022b. "'Stealth Predator': L.A.'s Famous Mountain Lion, P-22, Killed Hollywood Hills Chihuahua." *Los Angeles Times.* https://www.latimes.com/california/story/2022-11-21/big-cat-kills-chihuahua-in-hollywood-hills, retrieved February 27, 2023.

Spiegelman, Art. 2011. *The Complete Maus.* Pantheon Books.

Stoltz, Dustin S., and Marshall A. Taylor. 2021. "Cultural Cartography with Word Embeddings." *Poetics* 88:101567. doi: 10.1016/j.poetic.2021.101567.

Sullivan, R. 2005. *Rats: Observations on the History and Habitat of the City's Most Unwanted Inhabitants.* Perfection Learning Corporation.

Tavory, Iddo. 2010. "Of Yarmulkes and Categories: Delegating Boundaries and the Phenomenology of Interactional Expectation." *Theory and Society* 39(1):49–68.

Taylor, Marshall A., Dustin S. Stoltz, and Terence E. McDonnell. 2019. "Binding Significance to Form: Cultural Objects, Neural Binding, and Cultural Change." *Poetics* 73:1–16. doi: 10.1016/j.poetic.2019.01.005.

Thrasher, Steven. 2022. *The Viral Underclass: The Human Toll When Inequality and Disease Collide*. Macmillan.

Tilly, Charles. 2004. "Social Boundary Mechanisms." *Philosophy of the Social Sciences* 34(2):211–36. doi: 10.1177/0048393103262551.

Treasury Board and Finance Office of Statistics and Information-Demography. 2017. *2016 Census of Canada: Visible Minorities*. Government of Alberta.

Tsing, Anna Lowenhaupt. 2015. *The Mushroom at the End of the World: On the Possibility of Life in Capitalist Ruins*. Princeton University Press.

Unckless, Robert L., Philipp W. Messer, Tim Connallon, and Andrew G. Clark. 2015. "Modeling the Manipulation of Natural Populations by the Mutagenic Chain Reaction." *Genetics* 201(2):425–31. doi: 10.1534/genetics.115.177592.

United States Holocaust Memorial Museum. n.d. "Der Ewige Jude." *Der Ewige Jude*. https://encyclopedia.ushmm.org/content/en/article/der-ewige-jude, retrieved June 20, 2023.

Urry, John. 2007. *Mobilities*. Polity.

Vadillo, Monica Ann Walker. 2018. "From the Bestiary to the Margins: Animal Lessons on Sexuality and Motherhood in the Book of Hours of Charlotte of Savoy." In *Els Animals a l'Edat Mitjana*, 81–102.

Vaz, Ana S., Christoph Kueffer, Christian A. Kull, David M. Richardson, Joana R. Vicente, Ingolf Kühn, Matthias Schröter, Jennifer Hauck, Aletta Bonn, and João P. Honrado. 2017. "Integrating Ecosystem Services and Disservices: Insights from Plant Invasions." *Ecosystem Services* 23:94–107. doi: 10.1016/j.ecoser.2016.11.017.

Voyles, Traci Brynne. 2015. *Wastelanding: Legacies of Uranium Mining in Navajo Country*. University of Minnesota Press.

Wahlsten, Douglas. 1997. "Leilani Muir versus the Philosopher King: Eugenics on Trial in Alberta." *Genetica* 99(2):185–98. doi: 10.1007/BF02259522.

Warth, Gary. 2017. "Hundreds of Homeless People Kicked off the Streets as San Diego Battles Hepatitis A Outbreak That Has Killed 17." *Los Angeles Times*. https://www.latimes.com/local/lanow/la-me-san-diego-homeless-20170928-story.html, retrieved July 10, 2023.

Weindling, Paul. 1994. "The Uses and Abuses of Biological Technologies: Zyklon B and Gas Disinfestation between the First World War and the Holocaust." *History and Technology* 11(2):291–98. doi: 10.1080/07341519408581687.

Wekesa, J. Wakoli, Kimberly Nelson, Angela Brisco, Melvin Cook, and Kenn Fujioka. 2016. "History of Flea-Borne Typhus in Los Angeles County, California." *Proceedings and Papers of the Mosquito and Vector Control Association of California* 84:1–7.

Wickham, Hadley. 2022. "Rvest: Easily Harvest (Scrape) Web Pages." https://rvest.tidyverse.org/, retrieved July 11, 2023.

Wigglesworth, Alex. 2019. "Deputy Retaliated against Activist Who Protested Clearing of Homeless Encampment, Lawsuit Claims." *Los Angeles Times*. https://www.latimes.com/california/story/2019-08-13/deputy-retaliated-against-advocate-who-protested-clearing-of-homeless-encampment-lawsuit-claims, retrieved July 10, 2023.

Wohl, Hannah. 2015. "Community Sense: The Cohesive Power of Aesthetic Judgment." *Sociological Theory* 33(4):299–326. doi: 10.1177/0735275115617800.

Zahniser, David. 2019. "Lawyer Files $5-Million Claim, Saying L.A. City Hall Rat Problem Caused Her Illness." *Los Angeles Times*. https://www.latimes.com/local/lanow/la-me-ln-city-attorney-rat-flea-typhus-legal-claim-20190421-story.html, retrieved February 27, 2023.

INDEX

Aberhart, William, 67
agriculture, 71, 133, 175, 183
AIDS, 49, 140
Alberta, 1-2, 4-5, 16, 18-19, 23-24, 31-39, 41-75; Alberta Agenda, 36, 37, 72; borders of, 1-2, 4, 18-19, 23-24, 31-32, 34-35, 38-39, 41-45, 47, 49-50, 57-58, 60, 66, 71, 73, 186; Calgary, 19, 31, 32, 44, 46, 52, 53, 54, 64, 65, 67, 186; collective identity of, 19, 23, 24, 32, 33-38, 40, 43, 45, 50, 52-55, 57, 59, 61, 63-64, 66-67, 69-71, 73-74, 183, 185; Deerhome, 68; Department of Agriculture, 38, 46, 56, 57, 58; Department of Health, 38, 57, 58, 59, 186; Edmonton, 19, 65, 6; Eugenics Board, 67-68; Golden Jubilee, 61-63, 183; homesteaders in, 62-64, 69; indigenous people of, 69-70; Lethbridge, 65; as "rat-free," 1, 2, 4, 17-18, 23-24, 31-33, 35, 40-45, 47-49, 52-54, 56-58, 66, 70-71, 73-75, 86, 122, 181, 183, 185-86; Sexual Sterilization Act, 67-68, 70; 310-RATS, 18, 46, 48, 71
Alberta Department of Agriculture, 38, 46, 56, 57, 58
Alsask, Saskatchewan, 57
Angelo, Hillary, 9, 83, 193, 203

animals. *See* companion animals; non-human species; pets; *and specific species*
Anthropocene, 174-77, 189
anticommunism, 64, 66, 70
anti-immigrant sentiment, 24, 69, 72. *See also* immigrants and immigration
Apocalypse Now (film), 153
Apple Inc., 105
Army Corps of Engineers, US, 82, 118
Avenida Charles Darwin, Galápagos, 134

Baltra, Galápagos, 129
Bargheer, Stefan, 9, 10, 155, 203
Bartram, Robin, 116, 203
Beagle, HMS, 132, 135, 136
Benjamin, Ruha, 172-74, 204
Biehler, Dawn, 87, 204
biodiversity, 134, 139, 146, 150, 176
blue-footed booby, 142
Blue Marble, The (photograph), 7
borders. *See under* Alberta
boundary work, 4, 23, 31, 33, 38-39, 49-50, 53-55, 71, 73-75, 183
Bourdieu, Pierre, 15, 204
Brian (species eradication expert), 144-45, 158-61, 167, 169, 179-81
British Columbia, 2, 32, 39, 48
Buscaino, Joe, 100-101

Calgary, Alberta, 19, 31, 32, 44, 46, 52, 53, 54, 64, 65, 67, 186
Calgary Herald, 63
Campbell, Karl, 154, 204
Canada, 4, 16, 18, 19, 23, 32, 33, 35–37, 39, 51–52, 55, 61, 65, 68–69, 71, 72, 74, 179, 183, 185, 195, 196; government of, 35–37. *See also specific cities and provinces*
Canadian Shield, 38, 62, 147
carbon emissions, 175, 190
Carmen (environmental lawyer), 148, 150
Carolina (environmental lawyer), 158, 160, 162
Carson, Rachel, 6
Cascade Mountains, 2
Castro, Julian, 111
cats, 3, 14, 84, 133–34, 150, 165, 167, 182; feral, 89–90, 127, 131, 134, 158, 162
cattle, 31, 32, 42, 62, 133, 148
Chakrabarty, Dipesh, 175, 205
Channel Islands, California, 159. *See also specific islands*
charismatic wildlife, 4, 74, 84, 90, 97, 144, 193
Charles Darwin Research Center, Galápagos, 134, 138, 145, 155
Christianity, 24, 63–66, 69
climate change, 27, 97, 156, 176, 177, 199
Cohen, Stuart, 88
collective identity. *See under* Alberta
communism, 24, 64, 66, 70, 71, 186. *See also* anticommunism
companion animals, 5, 9, 10, 74, 89. *See also* pets
computational text analysis, 16, 18, 102, 107, 193, 197
conservation, 4, 14, 17, 21–23, 26–27. 97, 124, 127–34, 136–39, 141–44, 146, 148–49, 151–58, 160–64, 166–67, 169–74, 179–81, 185, 188–89. *See also* environmentalism
COVID-19, 22, 26, 182, 193, 197
CRISPR, 170–77, 179, 181, 189
Cresswell, Tim, 35, 205
Cronon, William, 136–37, 205
Crutzen, Paul, 174–75, 205

Darwin, Charles, 21, 26, 130, 132–35, 145
Davis, Mike, 82, 187
Deerhome, Alberta, 68
demographics, 14, 54, 55, 71, 106
Department of Health (Alberta). *See under* Alberta
Department of Sanitation (Los Angeles). *See under* Los Angeles
diversity, 55
dogs, 3, 85, 89, 90
Dojiri, Mas, 91–96
Domínguez Rubio, Fernando, 13, 49, 115, 140, 141, 151, 193, 206
Do My Own Pest Control, 1, 3
Douglas, Mary, 35, 115, 206
Douglas, Thomas, 172, 206
Downey, Robert, Jr., 112
Downtown Los Angeles Alliance for Human Rights, 104
Durkheim, Émile, 7–8, 206

Earth Day, 7
ecological meaning, 27, 129, 139, 141, 143, 146, 151, 177–78, 185, 189
ecology, 27, 86, 95–96, 121, 128–30, 133, 139–42, 151, 156–57, 163, 174, 177, 184–85, 188
ecosystem services, 139, 189
Ecuador, 4, 132, 136, 138, 153, 161, 162, 188; constitution of, 161–62. *See also* Galápagos Islands
Edmonton, Alberta, 19, 65, 68
El Niño, 118
Employment Equity Act (Canada), 55
encampments. *See under* Los Angeles
endangered species, 15, 161–63, 171, 174, 176–77
endemic species, 133–34, 138, 143–44, 153–54, 157, 169, 174, 176–77. *See also* native species
entomology, 58, 59, 71, 74, 186
Environics, 72, 196
environmental change, 5, 22, 27, 97, 130, 142, 146–49, 151, 169
environmentalism, 97, 130, 133, 136, 142, 143, 149, 154–56, 189
environmental justice, 5, 190
epidemiology, 89, 105, 123, 182, 187

Eternal Jew, The (Nazi propaganda film), 14
ethnic and racial others, 14, 70
eugenics, 67–70, 72, 75, 173
Eugenics Board (Alberta), 67–68
euthanasia, 85, 89
evolution, 21, 26–27, 130, 135, 145–47, 150, 161, 169, 171, 175–76, 185, 189
extermination. *See* rat extermination

feral cats. *See under* cats
"Firewall Letter" (Alberta), 37
fleas, 80, 88–89, 92–94
Floreana Island, Galápagos. *See under* Galápagos Islands
fortress conservation, 137, 155
fossil fuel industry, 37, 137, 139
Fund for Animals, 158, 167

Galápagos Conservancy. *See under* Galápagos Islands
Galápagos Islands, 4, 5, 16, 21, 22, 26, 27, 122, 127, 129–36, 138–39, 142–45, 149, 151, 153, 179, 182, 184, 188; Avenida Charles Darwin, 134; Baltra, 129; centipede, 144; Charles Darwin Research Center, 134, 138, 145, 155; Floreana Island, 22, 26, 127–28, 130–31, 139, 142–43, 147–50, 153, 156, 158, 171, 179, 188; Galápagos Conservancy, 22, 131, 153–54, 164; Galápagos National Park, 21, 127, 131, 134, 136–38, 148, 150, 153; North Seymour Island, 143; Perry Isthmus, 147, 149; Project Floreana, 22, 131, 153, 156, 158, 188; Project Isabela, 21, 131, 143, 147, 150, 154–56; Project Pinzón, 150; Puerto Ayora, 127, 134, 142; Puerto Velasco Ibarra, 127; tortoise, 21, 128–29, 131–35, 138, 143–46, 153–55, 164, 169, 177, 182, 185, 189; as UNESCO World Heritage Site and Marine Biosphere Reserve, 136–37
Galápagos National Park. *See under* Galápagos Islands
Garcetti, Eric, 110

Genetic Biocontrol of Invasive Rodents (GBIRd), 170
genetic editing, 27, 157, 170–74, 176, 179–180
Germany, 14, 50, 68
Ghana, 49, 140
Global News (Canada), 52
global rat distribution, 1
GMM models, 108, 200
goats, 21, 112, 131, 133–34, 144, 148, 150, 153–54, 159–60, 164, 167, 171
Golden Jubilee (Alberta), 61–63, 183
Gong, Neil, 104, 193, 207
good vs. bad nature, 25, 27, 81, 82, 85–90, 96–97, 102–103, 117–18, 120, 162, 169, 173, 178, 184–89
Gough Island, Tristan da Cunha, 163
Great Depression, 62
Grekul, Jana, 68, 207
Griffith Park, Los Angeles, 83–84, 86, 97

Harper, Stephen, 36
Harvey, David, 34, 207
Hayden, Tom, 89
Hayden Act (California), 89
hepatitis A, 91–93
hierarchy, 5, 10, 11, 14–15, 25, 53, 70–71, 73–74, 172, 185–87
Hitler, Adolf, 14, 53
homelessness. *See under* Los Angeles
homesteaders, 62–64, 69
housing insecurity, 20, 101, 103–4, 106–7, 109, 117, 124, 187
human-nonhuman relationships, 41, 149. *See also* companion animals; pets
hybridity, 87

Idaho, 2
immigrants and immigration, 5, 24, 50–51, 53, 54, 68–73, 138, 195
indigenous Albertans, 69–70
indoors vs. outdoors, 4, 20, 25–26, 79–82, 86–87, 96–97, 102–6, 114–15, 117–18, 120, 122–24, 183–84, 187
inequality. *See* rats: and social inequality

infestations, 4, 19–21, 25, 32, 39, 42–45, 80, 82, 84–85, 87, 90–91, 94–96, 99, 101–3, 105, 112, 114, 118–23, 188
introduced species, 17, 22, 133, 143–46, 150, 154–55, 157–58, 161–65, 169–71
invasive species, 4–5, 10, 47, 129–131, 147, 151, 153, 156–57, 159–67, 170–71, 177, 179, 182; eradication of, 129–31, 147, 151, 153, 156–57, 159–67, 170–71, 177, 179
Island Conservation (NGO), 22, 127, 131, 143–44, 158, 162–64, 170, 179

Jesse (pest control officer), 31, 39–41, 43–45, 51–52, 72
Jews, 14
jobs vs. the environment, 139
Jorge (Island Conservation staff member), 127
Judas goats, 154, 172

Kevin (Island Conservation director), 150
keystone species, 159, 162

Lethbridge, Alberta, 65
Liberal Party (Canada), 36, 51
Lock, Margaret, 176, 209
Lonesome George (tortoise), 165
Lopez, Steve, 112
Los Angeles, 4, 16, 19–21, 25–26, 79, 82–86, 88, 90–91, 93, 96, 99–103, 106–10, 112, 118–19, 122–23, 128, 179, 182, 186–87, 198; City Hall, 4–5, 20, 25–26, 79–88, 90, 93–97, 99–105, 197, 112, 116–22, 124, 184, 187; Civic Center, 19–21, 25–26, 81, 85, 87, 96, 102, 106, 117–19, 187; Department of General Services (GSD) and Personnel, 20, 79, 90, 99; Department of Sanitation, 20, 21, 26, 90, 91, 102, 118, 120–23; encampments in, 4, 20, 25–26, 100–106, 110, 112–14, 116–18, 120, 122, 187–88; Griffith Park, 83–84, 86, 97; Health Department, 91–92; homelessness in, 20–21, 26, 80, 100–12, 114, 116–18, 121–23, 187, 198–99; Police Department, 20, 90, 100, 113, 119, 121; Zoo, 83, 85
Los Angeles River, 82, 187
Los Angeles Times, 20–21, 25, 83–84, 88, 90, 93–96, 101–2, 107–8, 111–12, 114, 122
Los Cedros, Ecuador, 161
Los Feliz, California, 83

MacMurchy, Helen, 69
Malibu, California, 84
Manning, Ernest, 19, 24, 60, 62–71, 183, 186
Manning, Preston, 61
Massey, Doreen, 34–35, 210
McDonnell, Terrence, 12, 13, 49, 140, 142, 193, 210
McLaren, Angus, 67, 69, 211
mental illness, 88, 111–12
Merrill, Phil, 31–32, 38–39, 44–48, 56, 58
missionaries, 62–63
Mitchell Injunction (US court outcome), 100–101, 104, 116
Mohr, John, 192, 197, 211
Mona Lisa (painting), 151
Montana, 32
moral issues in conservation, 9–12, 15, 24–27, 33, 35, 49–51, 63–71, 73–74, 111, 118, 129, 142–46, 149, 151–53, 155, 156–58, 160–63, 169, 171–72, 177–78, 180, 181, 183, 186
moral purity, 24, 67, 69–71, 74, 173, 186
mountain lions, 83–85, 97
Muir, John, 137
Muir, Leilani, 67
multi-sited ethnography, 3, 16–18

Nading, Alex, 88, 211
National Academies of Sciences (US), 173
National Energy Program (NEP; Canada), 36–37, 61
nationalism, 38, 53
National Gideon Convention (Canada), 64
National Park Service (US), 83–84

native species, 4, 17, 21–22, 27, 128, 131, 133, 138, 140, 143–47, 149, 151, 153, 155–56, 162–67, 171, 176, 180, 182
nativist politics, 24, 35, 50, 52, 69–70, 73–74
natural language processing, 18, 23, 164, 198, 200
nature/society divide, 4, 25–26, 82–83, 86–87, 97, 124, 147, 149, 184, 189
Navy, US, 159
Nazi Germany, 14, 68
negative keystone species, 159, 162
neo-Nazis, 52
network maps, 164–68
New York City, 5–6
New York Times, The, 50, 138
NIMBYism, 116
1914 Social Service Congress of Canada, 69
nonhuman species, 2–11, 13–16, 25, 28, 33, 41, 53, 73–75, 80–86, 97, 105, 118, 124, 129, 131, 133, 138–41, 147, 149, 155, 157, 162–63, 169, 173–75, 183–86, 188–190. *See also specific species*
North Seymour Island, Galápagos, 143
Northwest Territories, 32

Ontario, 37, 69
oppositional identification, 19, 23, 33, 35, 37–39, 52, 67
opossums, 88–90, 187
Orange County, California, 82, 116–17; Orange County Public Health, 92

Peanuts (comic strip), 3
Pellow, David, 14, 71, 185, 191, 211, 212
Penfold, Steven, 37, 212
People for the Ethical Treatment of Animals (PETA), 158, 167
Perry Isthmus, Galápagos, 147, 149
pest control officers (PCOs), 4, 19, 23, 31, 51, 56–57, 71, 75, 179
pest control vendors, 81
pests, 1, 3–5, 9–10, 13, 15, 17, 19–21, 23, 25, 31, 39, 47, 51, 56–57, 71, 75, 81, 86–87, 94, 99, 101, 154, 179
pet rats. *See* rats: as pets

pets, 3, 6, 13, 47–48, 97, 182
Pinta Island tortoise, 165
pioneers, 61–63, 70
"pizza rat," 6
Point Mugu State Park, Malibu, California, 84
poison, 1, 22, 27, 40, 57–58, 84–85, 91, 128, 131, 144, 153, 157–58, 170–171, 178–180
Pomona, California, 88, 90
power, 5, 8, 14–15, 21, 25, 27, 33, 41, 56–57, 61, 65, 67, 70, 73, 90, 96, 102–3, 105–6, 123, 178, 183, 185, 189, 193
prairie, 18, 23, 31–32, 34–35, 43–45, 62, 69, 128, 179, 186
Price, Jenny, 82, 212
Project Floreana, 22, 131, 153, 156, 158, 188
Project Isabela, 21, 131, 143, 147, 150, 154–56
Project Pinzón, 150
propaganda, 14, 18, 47, 58, 66, 70
public spaces, 26, 102, 105–7, 117, 187
Puerto Ayora, Galápagos, 127, 134, 142
Puerto Velasco Ibarra, Galápagos, 127
pumas, 83–85, 97

Queen Astrid (ship), 127

Ratatouille (film), 6
rat control, 1–6, 11–13, 15–21, 23–28, 31–33, 35, 37–49, 51–61, 66, 68, 70–75, 81–82, 86–87, 89–92, 94–97, 99–103, 119, 120, 122–24, 128–29, 164–65, 170, 179, 181, 183–87, 191
Rat Control Zones (RCZs; Alberta), 32, 41–43, 51, 57, 71, 86
"rat czar" (New York), 6
rat extermination, 3–7, 9, 11, 13–16, 23, 47, 53, 58, 123, 128–29, 153, 157, 164–65, 169, 179–82, 191; as conservation, 4–7, 129, 153, 157, 164, 165, 169; as economic policy, 4–7, 183; as public health concern, 4–7
rat-landers, 45
rats: black rat (*Rattus rattus*), 133; brown rat (*Rattus norvegicus*), 133; as cultural objects, 11–13, 15, 23–

rats: black rat (*continued*)
24, 26, 33–34, 39, 45–46, 48–50, 52, 55, 59, 81, 128, 130, 140–43, 169, 181–82, 185, 188; as pets, 3, 6, 13, 47–48; "pizza rat," 6; and social inequality, 5, 8, 13–16, 25, 49, 70–71, 75, 105, 115, 123–24, 172–73, 183, 185–87, 189–190; as symbolic, 2, 11–16, 19–20, 24, 26, 33, 35, 39, 42–43, 45, 49–50, 52–53, 63, 66, 70–71, 73–75, 80–82, 90, 92–93, 95–97, 112, 122–24, 184–189; "toilet rat," 2. *See also* rat control; rat extermination
Reform Party (Canada), 36, 61
Robinson map projection, 1
Romanticism, 136
rural areas, 4, 17, 31, 37, 39, 44, 51–52, 54–55, 72
Russell, Edmund, 6, 212
Ryan (pest control officer), 94–95

San Clemente Island, California, 159–60
San Diego, California, 117; San Diego Police, 110; San Diego Public Health, 92
San Gabriel Mountains, 82
Santa Ana Civic Center, 117
Santa Barbara, California, 1–2, 127, 192
Santa Cruz Island, California, 127
Santa Monica Mountains, 83
Saskatchewan, 2, 4, 18–19, 31–32, 34, 38–39, 41–45, 47, 49, 52, 57, 64, 71, 183, 186
Savulescu, Julian, 172, 206
Schulz, Charles, 3
semantic environment, 198
Sexual Sterilization Act (Canada), 67–68, 70
Seychelles, 160
Seymour Airport, Galápagos, 129
Skid Row, Los Angeles, 20, 91, 101–2, 112, 119
Social Credit Party (Canada), 61, 64, 67–69, 71
Social Gospel movement, 68

Society for the Prevention of Cruelty to Animals (Canada), 48
South Equatorial Current, 132
Southern Poverty Law Center, 53
Special Law (1998; Ecuador), 138
sterilization, 25, 67–71, 186
Stoermer, Eugene, 174–75, 205
synthetic biology, 157, 171–72
synthetic gene drive, 170–74, 176, 178–81, 189

taxonomy, 12, 47–49, 70, 164
Third Reich (Germany), 14
310-RATS, 18, 46, 48, 71
Tim Horton's, 37
"toilet rat," 2
tourism, 127, 133, 138–39
Tristan albatross, 163
Trudeau, Justin, 51–52
Trudeau, Pierre, 36
Trump, Donald, 50–51
typhus, 20, 25–26, 79–81, 87–96, 100–101, 103, 105, 112, 114, 123, 182, 184, 187–88

UNESCO, 136–37
Uniform Manifold Approximation and Projection (UMAP), 199
United Kingdom, 63
University of Calgary, 44
University of California, Davis, 88
urban areas, 7, 9, 17, 25, 51–52, 54–55, 65, 71–72, 75, 77, 80–91, 95–97, 102–6, 116–24, 183–84
urban greening, 81, 83, 86
urban nature, 7, 9, 17, 25, 52, 75, 77, 81–90, 96–97, 102–3, 105–6, 117–18, 120, 122–124, 183–84
US-Mexico border wall, 35, 51
USSR, 24, 65

Venice Beach, California, 105, 113
villains, symbolic, 19, 24, 33–35, 49, 53, 63, 66, 70, 73, 186–87

Wagner, Richard, 153
Washington (state), 2

Washington Post, The, 104
Wesson, Herb, Jr., 80, 87, 93, 95, 121
western alienation, 36–37, 39
white supremacists, 53, 71, 73, 186
Williams, Raymond, 34

Woodsworth, J. S., 68–69
word embeddings, 107–9, 111, 198–200
World Bank, 160

Zyklon B, 14